Lecture Notes in Mathematics 1567

Editors:
A. Dold, Heidelberg
B. Eckmann, Zürich
F. Takens, Groningen

Subseries:
Nankai Institute of Mathematics,
Tianjin, P. R. China (vol. 10)

Advisor:
S. S. Chern, B.-j. Jiang

R. L. Dobrushin S. Kusuoka

Statistical Mechanics and Fractals

Springer-Verlag
Berlin Heidelberg New York
London Paris Tokyo
Hong Kong Barcelona
Budapest

Authors

Roland Lvovich Dobrushin
Institute for Problems of Information Transmission
Ermolovoj 19
103051 Moscow, Russia

Shigeo Kusuoka
Research Institute of Mathematical Sciences
Kyoto University
606 Kyoto, Japan

New address:
Department of Mathematical Sciences
University of Tokyo
113 Tokyo, Japan

Mathematics Subject Classification (1991): 82BXX, 82B31, 60K35, 60J60

ISBN 3-540-57516-2 Springer-Verlag Berlin Heidelberg New York
ISBN 0-387-57516-2 Springer-Verlag New York Berlin Heidelberg

2146/3140-543210 - Printed on acid-free paper

FOREWORD

The Nankai Institute of Mathematics held a special Year in Probability and Statistics during the academic year 1988-1989. We had over 150 specialists, professors and graduate students, who participated in this Special Year from August 1988 to May 1989. More than twenty outstanding probabilists and statisticians from several countries were invited to give lectures and talks. This volume contains two lectures, one is written by Professor R. L. Dobrushin, and the other one by Professor S. Kusuoka.

We would like to express our gratitude to Professors Dobrushin and Kusuoka for their enthusiasm and cooperation.

Ze-Pei Jiang
Shi-Jian Yan
Ping Cheng
Rong Wu

TABLE OF CONTENTS

ON THE WAY TO THE MATHEMATICAL FOUNDATIONS OF STATISTICAL MECHANICS

R.L.DOBRUSHIN

§0. Introduction

When I was a student in Moscow University at the end of the forties, I had to attend some lectures on physics. I had at that time a deep impression that although the content was very interesting to me, the form seems rather formidable. I asked myself, "Why don't they distinguish definitions from implications?" "Do they really fail to understand the difference between the necessary and the sufficient conditions?" "How can they formulate statements for which we see evident counterexamples?" I hoped that if I should become a professor, it would be possible for me to give a course of lectures on physics at a logical level consistent with the standard set by modern mathematics.

Later I understood that I had been naive and that the situation is not so simple. In fact my professors were not very bad. The style of their lectures reflected the logical level of modern theoretical physics which contrasts sharply with the logical level of modern mathematics. However, this was not always so. In the last century mathematics and physics were almost united. Readers will easily recall the names of great scientists who made important contributions both to physics and to mathematics. At that time there were no essential differences between the styles of exposition in the two subjects. Mathematicians and physicists spoke the same language and understood each other.

At the beginning of this century physics and mathematics began to move in different directions. Mathematics was incorporating very exciting new ideas: set theory, measure theory, modern algebra, functional analysis, topology, etc. New and higher standards of mathematical rigor were developed and any purported mathematical result which did not conform to this standard was considered either as erroneous or at least as lying outside of mathematics. We have now a standard universally accepted language for modern mathematics.

Physics went along another path. The new exciting ideas of quantum physics, theory of relativity, statistical physics, etc. posed attractive problems which required urgent solutions. In the beginning the methods of classical mathematical analysis developed in the last century were enough for their purposes. Physicists did not know modern mathematics and often treated it as an abstract and useless game. I have heard, for example, that our great physicist, Landau, said that he could invent all the mathematics which he needed. Such a point of view was even fashionable among physicists during that period. Physicists did not want to waste time on the fussy mathematical details needed for a rigorous proof. They considered something to be "proved" by using an argument which to a mathematician was just a rough plan or an idea for a future proof.

As a result, mathematicians and physicists almost ceased to understand and even hear each other.

A particular example of this estrangement is provided by probability theory (including the theory of random processes) on the one hand and statistical mechanics on the other. Both subjects were developing actively during these years, but even though they are concerned essentially with the same questions they were isolated from each other. And, for the most part, practitioners of the two subjects almost forget each other's existence.

In the middle of this century the situation began to change with the impulse for change coming from both mathematics and physics. Many good mathematicians began to realize that although there were still open and difficult problems in traditional mathematical directions the main constructions had been completed, and it was time for mathematics to have an infusion of new fresh ideas and problems. They turned to physics for inspiration. On the other hand, the constructions of modern physics became more and more complex and abstract. Unexpectedly, modern mathematics found applications in modern physics. The ideas of modern algebra and differential topology became essential to relativity theory and the theory of quantum fields. The ideas of functional analysis are basic for quantum mechanics, etc.

Interacton between the theory of probability and statistical physics also began to develop very rapidly. I think the investigations of recent years reveal that from the mathematical point of view statistical physics can be considered as a branch of probability theory. Seemingly all the main ideas and problems of statistical physics can be formulated in probabilitic language. But, of course, only a small portion of the assertions in statistical physics can be proved now at a mathematically rigorous level.

One should not suppose that all physicists will adopt the standards of mathematical rigor in pursuing their studies. In their obvious anxiety for quick results they will not cease to neglect mathematical logic. However, beginning in the fifties a different discipline with great rigor was evolved; it is the new science of mathematical physics. This is not the earlier "mathematical physics" which, for the most part, constituted a chapter in partial differential equations; but it is a science which is distinguished from physics and mathematics and lies between them. This new mathematical physics uses the language and standards of modern mathematics in studying the problems of physics. There are many scientists trained both in physics and mathematics who now work in this area. They also take up the most important task of helping mathematicians and physicists understand the problems and results in their respective fields in terms of what is apprehensible to both of them. Aside from several journals specializing in mathematical physics, there is now an international organization separate from the traditional physics and mathematics organizations.

The mathematical statistical physics about which I will speak in my lectures here has to be considered as a branch of mathematical physics strongly connected with probability theory, and I will speak only about the classical statistical physics. Classical theory means that it does not use the notions of quantum mechanics. However, all the ideas of classical statistical physics have their analogue in quantum statistical physics. Sometimes one refers to the area of mathematics used to study quantum mechanics problems as the non-commutative probability theory.

Statistical physics is strongly connected with other important branches of physics.

Thus, quantum field theory, which unites quantum mechanics and relativistic theory, can also be transformed into probabilistic language. To do this it is necessary to analytically extend quantum field theory to the case of complex time parameter, and to consider the case of pure imaginary time. Then we obtain a probabilistic picture of the so-called Euclidean field theory. Its connection with statistical physics is the same as that of continuous time random processes with discrete time random processes. Statistical physics can be used as a discrete approximation to quantum field theory, but the continuous version is much more complicated. In Euclidean quantum field theory it is necessary to consider Markov random fields in which realizations are distributions (generalized functions). This is not surprising from the point of view of classical probability theory. Every probabilist knows that almost all trajectories of a Markov diffusion process are continuous but non-differentiable functions. In the multidimensional case studied in quantum field theory the realizations of the natural Markov field become even worse. This very interesting theme requires a special exposition which we will not give here.

The aim of these lectures is to give an exposition, at a mathematical level, of the foundations of classical statistical mechanics. It is not easy even now. As a result of mathematical investigations in recent years we can at least reformulate all of the main notions of statistical mechanics in the language of mathematical definitions. But from the point of view of mathematicians modern statistical physics is something like a mix of some continents of well-developed mathematical theories with islands of separate mathematical results, amid a sea of open problems and conjectures (Of course, most physicists think of conjectures as results). Each year more and more conjectures get transformed into theorems. But now the majority of mathematical papers are devoted to problems of equilibrium statistical physics. Progress in this domain has found a systematic exposition in book form (See, for example, [Sinai(1982)], [Georgii(1988)]). The problem of the foundation of statistical physics, including the foundation of equilibrium statistical physics, is in the realm of nonequilibrium statistical physics, and here we have only isolated islands of theorems in a sea of conjectures.

Nevertheless, it seems that we now see a plan, a way to construct an orderly theory. I will try to give the main mathematical definitions, and explain the physical ideas underlying these definitions. I will also formulate a lot of open mathematical problems. Many of them seem very difficult now. I will also formulate theorems whenever they exist, but I will rarely give nontrivial proofs, leaving proofs to be found in the references. As is usual with young branches of mathematics all the proofs in mathematical nonequilibrium statistical physics are very complex and involved. Usually, with the development of a branch of mathematics the proofs become simpler and shorter. Since this is not so yet in the area discussed here it is not possible to give systematic proofs on the scale of these lectures.

I hope that the publication of these lectures helps to stimulate mathematical investigations in this field, especialy in China where I see a lot of talented young mathematicians who are eager to work on new problems. I am very grateful to Prof. Chen Mu-fa and his colleagues Chen Dong-ching and Zheng Jun-li who wrote up my lectures and helped to prepare their final version. Without their invaluable and well-qualified help this text would have never been written. I am also grateful to the members of the Nankai Institute of Mathematics for their hospitality. Here in Tianjin I have a happy possibility to meet Prof. M.D.Donsker.* I am very grateful to him for translating this introduction from its original Russian-Chinese dialect into real English.

Tianjin, 1987 December.

*which passed away prematurely in 1991.

§1. Realizations of the Classical Fluid Model

In this lecture I will speak mainly about the classical fluid model where the dynamics of particles is governed by the laws of classical Newtonian dynamics. It is the most natural and best-known model of statistical physics. Of course, as many physical models are, it is only an approximation to reality. For example, it does not take into account the quantum effects.

We will assume for simplicity that all particles are similar, i.e. they are particles of the same substance. The generalization to the case of particles of several types is not so complex. Denote by $(q, v) \in \mathbb{R}^d \times \mathbb{R}^d$ the particle with position q and velocity v, where d is a positive integer. In classical physics, $d = 3$. But some other dimensions also have physical interest (Dimension $d = 1$ corresponds to the statistics of threads, dimension $d = 2$ corresponds to the statistics of surfaces, dimension $d = 4$ corresponds to the problems arising from quantum field). So we will suppose that d is arbitrary and will, as in modern physics, follow the change of situation in the dependence of d.

If we have N particles, we denote their configuration by

$$\omega = ((q_1, \nu_1), \cdots, (q_N, \nu_N)) \in \hat{\Omega}_N \triangleq (\mathbb{R}^d \times \mathbb{R}^d)^N.$$

The realization space with an arbitrary number of particles is defined as follows:

$$\hat{\Omega} = \cup_{N=0}^{\infty} \hat{\Omega}_N$$

where $\hat{\Omega}_0$ is an empty set of particles.

Finally, from the physical point of view, it is natural to treat particles as undistinguished ones. For any $\omega \in \hat{\Omega}, \omega = ((q_1, \nu_1), \cdots, (q_N, \nu_N))$ and $A \subset \mathbb{R}^d \times \mathbb{R}^d$, let

$$\pi_\omega(A) = |\{i \in \{1, \cdots, N\} : (q_i, \nu_i) \in A\}|$$

where $|A|$ = the number of elements in A, π_ω is an integer-valued measure on Borel σ-algebra in $\mathbb{R}^d \times \mathbb{R}^d$ and $\pi_\omega(\mathbb{R}^d \times \mathbb{R}^d) < \infty$. In this way, we have defined a mapping from $\hat{\Omega}$ into the space $\hat{\Pi}$ of such measures, we call $\pi \in \hat{\Pi}$ an ordinary realization if $\pi(x) = \pi(\{x\}) \leq 1$ for all $x \in \mathbb{R}^d \times \mathbb{R}^d$. Let

$$X_\pi = \{x \in \mathbb{R}^d \times \mathbb{R}^d : \pi(x) = 1\}.$$

So we can interpret the ordinary realization X_π as a finite subset of the space $\mathbb{R}^d \times \mathbb{R}^d$. For most of the situations it is enough to consider only ordinary realizations and we will do so in almost all our lectures.

Statistical physics studies a finite but very large system of particles. One of the main features of mathematical approach is the explicit consideration of an infinite particle system which makes many notions of statistical physics much more sharp and accurate. We will systematically use in these lectures such a point of view. Let

$\Pi \triangleq \{\pi : \pi$ is an integer-valued measure on $\mathbb{R}^d \times \mathbb{R}^d$ with $\pi(S \times \mathbb{R}^d) < \infty$ for all compact subsets $S \subset \mathbb{R}^d\}$

$\Omega \triangleq \{X \subset \mathbb{R}^d \times \mathbb{R}^d : |X \cap (S \times \mathbb{R}^d)| < \infty$ for all compact subsets $S \subset \mathbb{R}^d\}$,

We will call elements of Π and Ω locally finite realizations.

We assert that $\Omega \subset \Pi$ by using the identification similarly used above for a finite particle system.

Let $\mathcal{B}_\Pi$ be the smallest σ-algebra with respect to which the functions $f(\pi) = \pi(S \times \tilde{S})$, $\pi \in \Pi$ are measurable, where S and $\tilde{S}$ are compact subsets of $\mathbb{R}^d$. Let $\mathcal{B}_\Omega$ be the smallest σ-algebra with respect to which the functions $f(X) = X \cap (S \times \tilde{S})$, $X \subset \Omega$ are measurable for all pairs $(S, \tilde{S})$ of compact subsets of $\mathbb{R}^d$. It is easy to prove the following fact:

1) $\Omega \subset \Pi$ is a measurable subset of Π, and the restriction of $\mathcal{B}_\Pi$ on Ω coincides with $\mathcal{B}_\Omega$.

We will leave the proof of this fact to our reader as an exercise.

Given a compact subset $V \subset \mathbb{R}^d$, we define Π_V and Ω_V by replacing $\mathbb{R}^d \times \mathbb{R}^d$ with $V \times \mathbb{R}^d$ in the definitions of Π and Ω respectively. Similarly we introduce $\mathcal{B}_{\Pi_V}$ and $\mathcal{B}_{\Omega_V}$. Of course $\Omega_V \subset \Omega$, $\Pi_V \subset \Pi$ and it is easy to check that these embeddings are measurable.

Now we have two measurable spaces $(\Omega, \mathcal{B}_\Omega)$ and $(\Pi, \mathcal{B}_\Pi)$. We will construct a basic measure on them, connected with the usual Lebesgue measures in Euclidean space. Define $\hat{\Omega}_V = \cup_{N=0}^{\infty} \hat{\Omega}_V^N$ where $\hat{\Omega}_V^N = (V \times \mathbb{R}^d)^N$, and the transformation $\alpha : \hat{\Omega}_V \to \Pi_V$, $\alpha(\omega) = \pi_\omega(\cdot) \in \Pi_V, \omega \in \hat{\Omega}_V$.

Let $\hat{\lambda}_N$ be the Lebesgue measure on $(V \times \mathbb{R}^d)^N, N = 1, 2, \cdots$. Define a measure $\hat{\lambda}$ on $\hat{\Omega}_V$ such that $\hat{\lambda}(A) \triangleq \hat{\lambda}_N(A)$ for $N \geq 1$ and all measurable subsets $A \subset (V \times \mathbb{R}^d)^N$, and set

$$\Pi_V^N = \{\pi \in \Pi_V : \pi(V \times \mathbb{R}^d) = N\}, \Pi_V = \cup_{N=0}^{\infty} \Pi_V^N.$$

For $A \subset \Pi_V^N, N > 0$, we define

$$\lambda(A) \triangleq \frac{1}{N!} \hat{\lambda}(\alpha^{-1}(A))$$

and assume $\lambda(\Pi_V^0) = 1$ (The set Π_V^0 consists of a unique measure $\pi_V^0 \equiv 0$). For any $B \subset \Pi_V$, there is a partition of B, $B = \sum_{i=0}^{\infty} A^i, A^i \in \Pi_V^i, i = 0, 1, 2, \cdots$, so we can define

$$\lambda(B) \triangleq \sum_{i=0}^{\infty} \lambda(A^i).$$

For any compact subsets $V_1, V_2, V_1 \subset V_2 \subset \mathbb{R}^d \times \mathbb{R}^d$, the restriction of λ_{V_2} to V_1 is equal to λ_{V_1}. This is because of the consistency of Lebesgue measures. Since $\Pi = \cup_{V \subset \mathbb{R}^d} \Pi_V$, then using the previous property we can define a measure on Π which is also denoted by λ and will be called the basic measure on Π. By definition, we have

2). If $V_1, V_2 \subset \mathbb{R}^d$ are compact subsets and $V_1 \cap V_2 = \phi, V \triangleq V_1 \cup V_2$, then $\Pi_V = \Pi_{V_1} \times \Pi_{V_2}$ $\mathcal{B}_{\Pi_V} = \mathcal{B}_{\Pi_{V_1}} \times \mathcal{B}_{\Pi_{V_2}}$ and $\lambda_V = \lambda_{V_1} \times \lambda_{V_2}$.

3). $\lambda(\Pi \setminus \Omega) = 0$.

The proof of these facts can also be considered as an exercise. Very often we will treat the basic measure as a measure on the space of ordinary realizations Ω. The reader who knows well the probability theory understand that λ is a Poisson measure well known in the theory of point random fields.

§2. Dynamics of a Finite System

Suppose that we are given N particles $\omega = ((q_1,\nu_1),\cdots,(q_N,\nu_N))$ and an interacting potential U. Here we consider only the pair potentials which are translation invariant, isotropic, and so we interpret a potential as a function on $\mathbb{R}^+ = \{x \in \mathbb{R} : 0 < x < \infty\}$ into $\mathbb{R}$. We will consider the following equations of motion of Newtonian type.*

$$\text{(2.1)} \qquad \begin{cases} \frac{dq_i(t)}{dt} = \nu_i(t) & i = 1,\cdots,N, t \in (0,\infty) \\ \frac{d\nu_i(t)}{dt} = -m\ grad_{q_i} \sum_{j=1, j\neq i}^{N} U(|q_j - q_i|), \end{cases}$$

where m is the mass of one particle. If we denote the momentum by $p_i = m\nu_i$, we have the following Hamiltonian equations:

$$\begin{cases} \frac{dq_i}{dt} = grad_{p_i} H(q_1,\cdots,q_N,p_1,\cdots,p_N) \\ \frac{dp_i}{dt} = -grad_{q_i} H(q_1,\cdots,q_N,p_1,\cdots,p_N) \end{cases} \qquad i = 1,\cdots,N.$$

where

$$H(q_1,\cdots,q_N,p_1,\cdots,p_N)$$
$$= \frac{1}{2}\sum_{i=1}^{N} \frac{(p_i)^2}{m} + \sum_{\substack{i,j=1 \\ i\neq j}}^{N} U(|q_i - q_j|).$$

This last quantity is called the Hamiltonian of the system. In the following we let $m = 1$.

Mathematicians often ask: What function U is a real physical potential? The question is not correct. First of all, any classical model is only a rough approximation to a quantum model, and our choice of potential is such an approximation in some sense. Secondly, there are a lot of types of particles, and different types of potentials are naturally for different types of particles. Finally, it is better to have results for some potential than to have no results. So potentials having the simplest analytical structures are often considered. But the results for any potentials are interesting. It is especially interesting to have results applicable to a wider class of potentials and to follow the change of qualitative property of the system in the dependence on the potential. This conclusion should sound pleasant to mathematicians.

Now we will give some typical examples of the potentials.

1. Lenard-Jons potential

$$U(x) = \begin{cases} \frac{K_1}{|x|^n} - \frac{K_2}{|x|^l}, & \text{if } x \neq 0 \\ \infty, & \text{if } x = 0, \end{cases}$$

where K_1 and K_2 are positive constant, l and n are positive integers. In the 3-dimensional case, it is often to suppose that

$$l = 6, \qquad n = 2l = 12.$$

Figure 1 indicates one of this kind of potentials.

* See for example [Arnold(1978)] in connection with elementary notions of mechanics used in these lectures.

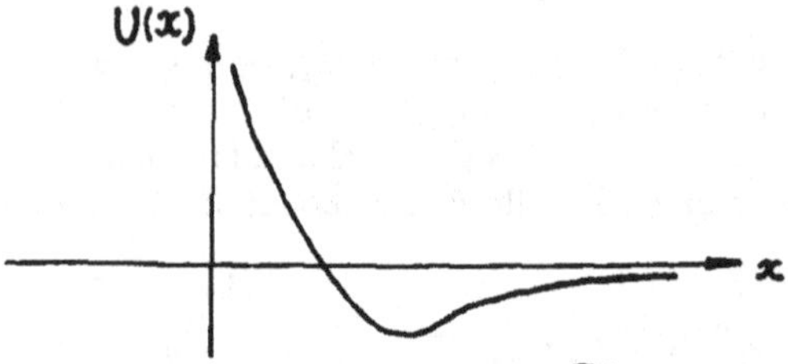

Figure 1

This structure of the potential can be justified by some quantum type of consideration for the case of one-atom gas. The decreasing part of the graph corresponds to the repulsion of particles and the increasing part corresponds to their attraction. The value $U(0) = \infty$ means that two particles can not collide with each other.

2. Morse potential

$$U(x) = K[1 - \exp\{-\alpha(x - \hat{x})^2\}]^2, \quad x \geq 0,$$

where α and K are constants; $\hat{x}$ is a fixed reference point.

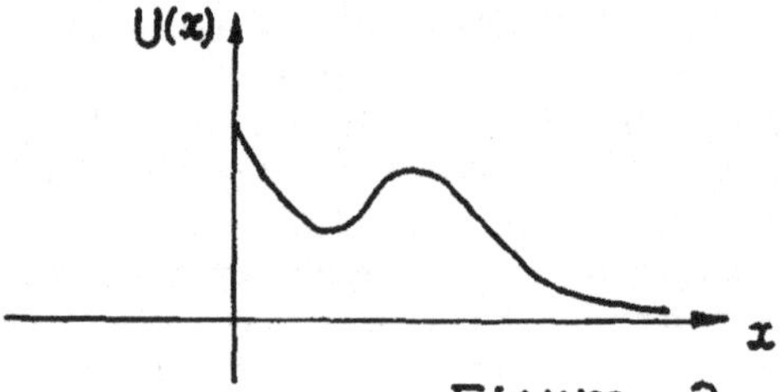

Figure 2

Such a potential is used for two-atom particles. The value $U(0) < \infty$ means that the two particles can meet together. Of course this is not very natural from the physical point of view.

3. Hard core potential

Suppose that there exists an $r > 0$ such that

$$U(x) = \infty, \qquad |x| \leq r;$$

or, in another language, each particle is a hard sphere of diameter r. It means that particles can not be closer to one another than at a distance r.

Here we give three potentials of this kind.

3(a). $U(x) \to \infty$ as $x \to r$:

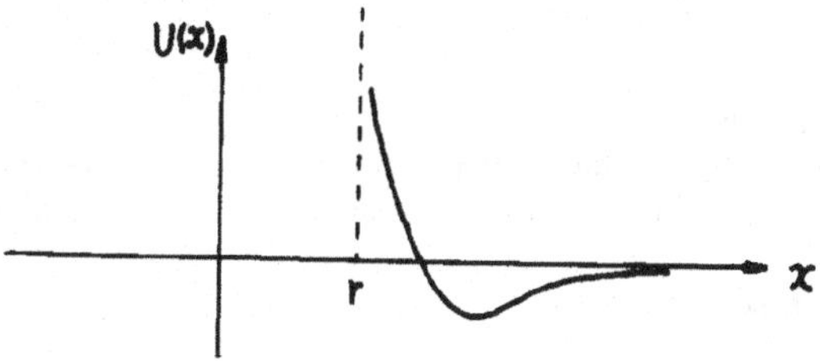

In this case particles can not collide because of repulsion. When the distance between them is close to r there is a very strong repulsion.

3(b). $U(x) \to c \neq \infty$ as $x \to r$:

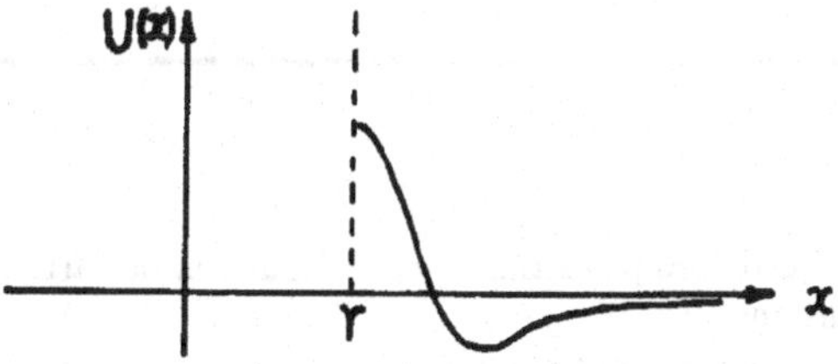

In this case particles can collide. So we have to add to the equations of motion (2.1) some boundary conditions.
Usually we will suppose that the collision is elastic one (As for ordinary billiard-balls, see later). To have only pair collisions will permit us to define a unique solution to the equations of motion. This is not so in the case of multiple collisions (See graph below).

Pair collision

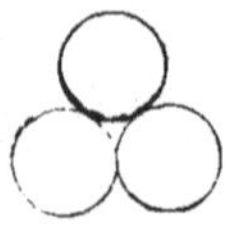

Collision of three particles.

It is natural to suppose that under ordinary circumstances the multiple collisions occur with probability 0, but this question has not been thoroughly studied at a mathematical level.

3(c). $U(x) \equiv 0$, $|x| > r$:

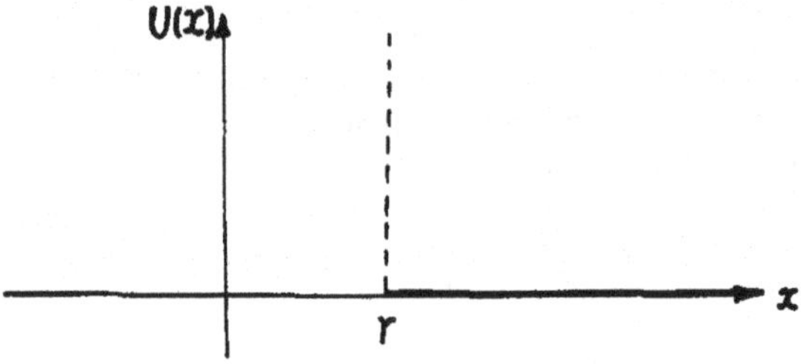

It is the case of pure hard core potential where interaction of particles arises only at the moment of their collisions.

Generally we will suppose that $r = 0$ is a possible value for the hard core diameter. The case

$$U(x) \equiv 0, x \in \mathbb{R}^+$$

corresponds to that of an ideal gas where particles do not interact.

We shall mainly suppose that some conditions of smoothness are true:

1^0 The smooth potential. The potential $U(x)$ has continuous first derivative in $x \in (0, \infty)$. In this case we can find a unique solution to the equations of motion under certain initial conditions.

2^0 The smooth hard core potential, i.e.

$$U(x) = \infty, \quad |x| \leq r,$$
$$U(x) \to \infty, \quad \text{as } x \to r,$$

and $U(x)$ has continuous first derivative for $x \in (r, \infty)$. This means that when particles get closer and closer, the energy becomes very great. By using the law of conservation of energy we can again prove the existence of a unique solution.

Now we give some notations. In the following we consider only ordinary realization, and in an obvious way redefine the motion as a motion of nondistinquished particles.

Let $\mathcal{A}$ denote the set of all finite subsets of $\mathbb{R}^d \times \mathbb{R}^d$. For each $t \in \mathbb{R}_+ = [0, \infty)$, we define a mapping $T_t : \mathcal{A} \to \mathcal{A}$ as follows: $T_t a$ is the realization at the moment t if $a \in \mathcal{A}$ is an initial realization, i.e. $T_t a = a(t)$, where $a(t)$ is the solution of the equations of motion with initial condition a. Using the well-known properties of the solution of differential equations, we can define also T_t for $t < 0$ and have $T_{t+s} = T_t \cdot T_s$ for all $s, t \in \mathbb{R}$. So $\{T_t : t \in \mathbb{R}\}$ is a group of transformations if, of course, restrictions discussed above are valid.

In order to describe the laws of conservation, we have to introduce additional functions. Let ϕ be a function on $\mathbb{R}^d \times \mathbb{R}^d$. For any $a \in \mathcal{A}$, we define $F : \mathcal{A} \to \mathbb{R}^d, (d = 1, 2, \cdots)$ by setting

$$F(a) = \sum_{(q,\nu)\in a} \phi(q, \nu).$$

We call this kind of F the (translation-invariant) first-order additive functional on the realization space.

We give some examples of the additive functional.

Example 1.

$$N(a) = \sum_{(q,\nu)\in a} 1, \quad \phi(\cdot) \equiv 1.$$

This is the total number of particles.

Example 2.

$$M(a) = \sum_{(q,\nu)\in a} \nu, \quad \phi(q, \nu) = \nu.$$

This is the momentum of the system.

When $F(a(t)) \equiv const.$ for any initial realization $a(0)$, we have a law of conservation. So, in Example 1, it is the law of conservation of particles (or of the mass); in Example 2, it is the law of conservation of momentum. These laws of conservation for the dynamical system of finite particles are well-known from elementary courses in mechanics.

If ϕ is defined on $(\mathbb{R}^d \times \mathbb{R}^d) \times (\mathbb{R}^d \times \mathbb{R}^d)$, then

$$F_2(a) = \sum_{(q,\nu_1),(q_2,\nu_2)\in a} \phi(q_1, \nu_1 ; q_2, \nu_2),$$

is called an additive functional of second order. We will call this functional translation-invariant if

$$\phi(q_1, \nu_1; q_2, \nu_2) = \phi(q_1 - q_2, \nu_1, \nu_2).$$

We can also define a translation-invariant additive functional of any order like this.

The well-known law of conservation of energy can be described by the following translation-invariant additive functional of second order.

$$E(a) = \frac{1}{2} \sum_{(q,\nu)\in a} \nu^2 + \sum_{(q_1,\nu_1),(q_2,\nu_2)\in a : q_1 \neq q_2} U(|q_1 - q_2|)$$

where U is the potential defined previously. For the case when collisions are possible it is well known from mechanics that the laws of conservation of momentum and of energy are true when the colliding of particles can be considered as elastic collisions.

There exist non-translation-invariant laws of conservation. For example, the law of conservation of central momentum

$$F(a) = \sum_{(q,\nu)\in a} [q, \nu],$$

where $[\cdot, \cdot]$ is the scalar product in Euclidean space. Such non-translation-invariant conservation laws are not essential for the problems concerning the foundation of statistical mechanics.

There are some degenerate systems for which we have a lot of additional translation-invariant laws of conservation. For example, this is the case if dimension $d = 1$ and

$$U(x) = c(shAx)^{-2};$$

and also the limiting case $A \to 0, c/A^2 \to const$ with the potential

$$U(x) = cx^{-2}.$$

Here we have an infinite system of non-trivial translation- invariant laws of conservation.

Another example gives the pure hard core potential for dimension $d = 1$. We call the corresponding system as one of 1-dimensional hard rods. Here at the moment of collision the two particles simply exchange their velocities. So for any function $\phi(\nu)$ the relation

$$F_\phi(a) = \sum_{(q,\nu)\in a} \phi(\nu)$$

gives a translation-invariant law of conservation.

An important hypothesis states that the desribed cases are only exceptional cases, and (may be under some mild additional hypothesis of a general type) for all other potentials there are no additional laws of conservation. Under some strong additional conditions about potentials and functionals this important hypothesis has been proved by [Gurevich, Suhov 1976, 1982].

The structure of additive functionals plays a very important role in the description of the structure of equilibrium states. In both degenerated models described above the last structure also has a special form, see §4 for hard rod system and see [Chulaevsky, 1983] for the system with potential cx^{-2}.

We need to introduce other important properties of finite particle systems. For any $A \in \mathcal{A}$, it follows from Liouville theorem that

$$\lambda(A) = \lambda(T_t A), \quad t \geq 0$$

where measure $\lambda(\cdot)$ was defined in §1. So we have a dynamic system with an invariant measure. The other property is time-reversibility. This means that

$$T_{-t}a = (T_t a^*)^*, \quad t \in \mathbf{R}^1, \quad a \in \mathcal{A},$$

where $a = \{(q, \nu)\} \to a^* = \{(q, -\nu)\}$.

Sometimes it is better to think that the motion of particles is bounded by some volume $V \subset \mathbb{R}^d$, where V is a closed bounded domain. To define the dynamics we can introduce three kinds of boundary conditions.

1^0 We change the equations of motion into

$$\begin{cases} \frac{dq_i}{dt} = \nu_i \\ \frac{d\nu_i}{dt} = -grad_{q_i}(\sum_{k\neq i} U(|q_k - q_i|) + \sum_i \bar{U}(q_i)) \end{cases}$$
$$i = 1, 2, \cdots, N,$$

where $\bar{U}(q)$ can be interpreted as an external field. We suppose that $\bar{U}$ is a smooth function satisfying the following condition: There exists a constant $r > 0$ such that

$$\bar{U}(q) \equiv 0, \quad \text{if } dist(q, V^c) > r$$
$$\bar{U}(q) \to \infty, \quad \text{as } dist(q, V^c) \to 0.$$

In this system the particles can not reach the boundary of V and so the dynamic system is well-defined. But the law of conservation of momentum is not true for such a system, and the law of conservation of energy will be true only if we define the energy as

$$E(a) = \frac{1}{2} \sum_{(q,\nu)\in a} \nu^2 + \sum_{(q_1,\nu_1),(q_2,\nu_2)\in a} U(|q_1 - q_2|)$$
$$+ \sum_{(q,\nu)\in a} \bar{U}(q).$$

2^0 Suppose that the particles can reach the boundary of a finite volume V and elastic collision with the boundary takes place, see Figure 2^0. This means that the particles are confined in a volume and the mass of the container is very large. In this system the law of conservation of energy is true in the original form. But the law of conservation of momentum is again not fulfilled.

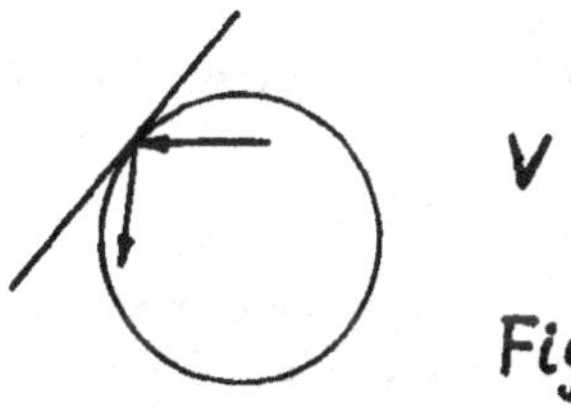

Figure 2°

3^0 The particles move on a torus and the equations of motion are defined on the torus in a natural way. This case is called the case of periodic boundary conditions. It has no physical interpretation, but both law of conservation of energy and law of conservation of momentum are true. Of course in all these three cases the law of conservation of the number of particles holds.

§3. Dynamics of an Infinite System

Let us begin to discuss the infinite particle systems. We will denote a realization $a(t) \in A$, $-\infty < t < \infty$ by

$$a = \{(q_1(t), \nu_1(t)), (q_2(t), \nu_2(t)), \cdots, (q_n(t), \nu_n(t)), \cdots\}.$$

For simplicity we consider only a finite range of potentials, that is, $U(r) \equiv 0$, if $|r| > r_0$, where r_0 is called the radius of interaction. The infinite motion equation is

$$\begin{cases} \frac{dq_i(t)}{dt} = \nu_i(t) & i = 1, 2, 3, \cdots, \\ \frac{d\nu_i(t)}{dt} = -grad_{q_i} \sum_{j \neq j'}^{\infty} U(|q_j(t) - q_{j'}(t)|) \end{cases}$$

and we have to discuss a question about the existence and the uniqueness of its solution.

From now on, let $\Omega_n = \{q \in \mathbb{R}^d : |q| \leq n\}$ and for all $a \subset \mathbb{R}^d \times \mathbb{R}^d$,

$$a_n = a|_{\Omega_n} = \{(q, \nu) \in a : |q| \leq n\}$$

$$a_n(t) = T_t a_n.$$

Because a_n is finite, the variables $a_n(t)$ are well defined. But we will study the limit

$$a(t) = \lim_{n \to \infty} a_n(t).$$

First we must introduce a topological structure for infinite realizations so that the above limit has a meaning. We begin with the case of finite realizations. If $a^1, \cdots, a^k, \cdots \in \mathcal{A}$, $a \in \mathcal{A}$, we define

$$a^k \mapsto a = \{(q_1, \nu_1), \cdots, (q_N, \nu_N)\}$$

if and only if the following two conditions are satisfied:

i) There exists a $k_0 > 0$ such that $N(a^k) = N(a)$, for $k \geq k_0$

ii) For any $\varepsilon > 0$, there exists an $n_0 \geq 0$ such that for $n \geq n_0$ we have

$$a^n \cap B((q_i, \nu_i), \varepsilon) \neq \ominus, i = 1, 2, \cdots, N$$

where $B(x, \varepsilon)$ is a ball in $\mathbb{R}^d \times \mathbb{R}^d$ with center x and radius ε, and $\ominus$ is an empty set.

Now we can define the convergence in infinite systems. Let $a^k \in \Omega, k = 1, 2, \cdots$. We say that a^k converges to $a \in \Omega$, if for any open bounded subset $Q \subset \mathbb{R}^d$,

$$a^k_Q \to a_Q, \quad \text{as } k \to \infty,$$

where $a_Q = \{(q, \nu) \in a; q \in Q\}$. We will always consider this topology in the space of realizations. The following example shows that the limit $\{a_n(t), n \geq 0\}$ will not always exist for every initial configuration $a \in \Omega$.

Example. Let $U(q) \equiv 0$ i.e. we cousider an ideal gas. Then $\nu(t) \equiv \nu = const.$ and $q(t) = q(0) + vt$. If we choose an initial configuration

$$a = \{(1,-1),(2,-2),\cdots,(m,-m),\cdots\},$$

then

$$a_n(1) = \{0,-1),(0,-2),\cdots,(0,-n)\}.$$

and so the limit of $a_n(1), n \to \infty$ don't exist. But if we suppose that for $a = \{(q,\nu)\}$

$$\frac{|\nu|}{|q|} \to 0, \quad \text{as } q \to \infty,$$

the limit is well defined. This is because

$$q + \nu t = q(1 + \frac{\nu}{q} \cdot t) \sim q, \ (\text{as } q \to \infty).$$

Similar difficulties arise for other more complex potentials.

So we have to find a measurable $\mathcal{N} \subset \Omega$ such that the following seven conditions are satisfied:

1) $\lim_{n\to\infty} a_n(t)$ exists for $a \in \mathcal{N}$. Let $T_t a = a(t) = \lim_{n\to\infty} a_n(t)$, if $a(0) = a$;

2) $T_t\mathcal{N} \subset \mathcal{N}$;

3) $T_s \cdot T_t = T_{s+t}$, $s,t \in \mathbb{R}^1$;

4) The limit $a(t)$ is the unique solution of the infinite system described above.

5) The infinite dynamical system is time reversible, i.e., $T_{-t}a = (T_t a^*)^*, a \in \mathcal{N}$, where $a = \{(q,\nu)\} \mapsto a^* = \{(q,-\nu)\}$;

6) The infinite dynamical system is λ-invariant in $\mathcal{N}$, i.e., $\lambda(A) = \lambda(T_t A), A \subset \mathcal{N}$;

7) The three main laws of conservation hold.

The last statement needs an interpretation. If $a \in \mathcal{A}$, then conservation means that $F(a(t)) \equiv constant$. But what is a meaning of conservation for $a \in \Omega$? If the limit

$$\lim_{n\to\infty} \frac{F(a_n)}{|\Omega_n|} = \bar{F}(a)$$

exists, we call $\bar{F}(\cdot)$ a density of a functional F. If for any $a(0) \in \mathcal{N}$ the existence of $\bar{F}(a(0))$ implies the existence of $\bar{F}(a(t))$ and the equality

$$\bar{F}(a(t)) \equiv \bar{F}(a(0)),$$

then we say that the infinite dynamical system has a law of conservation with respect to the functional F.

It is also natural to expect that if we define $a_n(t)$ by the help of dynamics in a finite volume with some boundary conditions then the limit will also exist and equals $a(t)$.

It is also possible to define $a_n(t)$ by replacing the restriction $a_n(0)$ of $a(0)$ on a cube Ω_n by its restrictions on a more general sequence of volumes. It is only necessary to suppose that the corresponding volumes of the sequence go to infinity in Van-Hove sense. We say that the volumes of the sequence $V_n \subset \mathbb{R}^d$ go to infinity in Van-Hove sense if

1) The Lebesgue measure $|V_n| \to \infty$.

2) For any $\tau > 0$ and $\partial_\tau V = \{x \in V : dist(x, V^c) \leq \tau\}$,

$$\lim_{n\to\infty} \frac{|\partial_\tau V_n|}{|V_n|} = 0.$$

The main difficulty of finding a solution of an infinite dynamical system is to prove the compactness of the sequence of $a_n(t)$ for a fixed $t > 0$. In our topology we have the following

Proposition 1. $\mathcal{A}' \subset \Omega$ is compact if and only if for any open $Q \subset \mathbb{R}^d$, the following two conditions are satisfied for certain constants $C_Q^1, C_Q^2 > 0$ and any $a \in \mathcal{A}'$:

i) $|a_Q| \leq C_Q^1 < \infty$,

ii) $|\nu| \leq C_Q^2$, if $(q, \nu) \in a_Q$.

The proof is left as an exercise.

Whenever these conditons of compactness have been checked it is not difficult to prove all the statements formulated above. But checking the compactness is really a difficult problem, which has been solved only partially up to now. We try further to describe the general features of the results and to explain. the idea of their proofs and the main difficulties.

The study of this problem was initiated by Lanford (1968-1969). He considered the one-dimensional case ($d = 1$) and supposed that the potential $U(x)$ is a smooth function with a finite range (It means $U(x) = U(|x|)$ has continuous derivative in $\mathbb{R}^1$ and $U(x) \equiv 0$ for $|x| \geq \tau_0$ for some $\tau_0 > 0$). This implies that $U(0) < \infty$ what seems not very natural from the physical point of view because it means that the two particles can go through the same position. Lanford proved that the infinite particle dynamical system exists (that is, $\lim_{n\to\infty} a_n(t) = a(t)$ exists for all t) under the following two conditions about the initial configuration:

1) $\sup_{(q,\nu)\in a} |\nu|/(\ln_+ |q|) < \infty$

2) $\sup_{y\in\mathbb{R}} |a_{[y-\ln_+|y|, y+\ln_+|y|]}|/2(\ln_+ |y|) < \infty$

where $\ln_+ |q| = \max(1, \ln|q|)$ and $|a_V|$ is the number of a in interval V.

If we suppose that for any moment $t \geq 0$ the condition $i)$ of Proposition 1 is true, then the boundedness of the potential and the equations of motion imply that the derivatives $\frac{dv_i}{dt}$ are also bounded. Then, the condition 1) about the initial realization will imply the boundedness of the velocities (See 2)) at moment $t > 0$. On the other side, if we know that the velocities are bounded then the condition 2) about the initial realization will imply that the number of particles in the volume at moment t (See 1)) is bounded. Of course these two notes generate a closed logical ring. They can be used for the mathematical proof if we follow the change of constants C_Q^1, C_Q^2 in Proposition 1 in the evolution of time. It is possible to say that Lanford's proof is founded on the law of conservation of the number of particles.

Dobrushin and Fritz (1977) and Fritz and Dobrushin [1977] gave other conditions which are different from those of Lanford's. They considered dimension $d = 1$ or 2 and potential $U(x)$ with a finite range. Their conditions are as follows:

1) $|x| \cdot |gradU(x)| \leq a + b \cdot U(x)$ for a small enough x, and some $a < \infty, b < \infty$

2) Stability conditon: There exists two constants $A \geq 0$ and $B > 0$ such that

$$\sum_{j,k=1, j\neq k}^{n} U(|q_j - q_k|) \geq -A \cdot n + B \cdot |\{(i,j) : |q_i - q_j| < 1\}|$$

holds for any set $q_1, \cdots, q_n$ of points in $\mathbb{R}^d$.

3) Condition about the initial configuration:

$$H = \sup_{\mu} \sup_{\sigma > \log_+ |\mu|} H(\mu, \sigma, a) \cdot \sigma^{-d} < \infty$$

where

$$H(\mu, \sigma, a) = \sum_{\substack{|q_i - \mu| < \sigma \\ (q_i, v_i) \in a}} (|v_i|^2 + A) + \Big| \sum_{\substack{|q_i - \mu| < \sigma \\ (q_i, v_i) \in a}} \sum_{j \neq i} U(|q_j - q_i|) \Big|$$

$\sigma > 0, \mu \in \mathbb{R}^d, a \in \Omega; A \geq 0$ being the same as in 2).

An example of the potential which satisfies the conditions 1) and 2) is

$$U(x) \sim c \cdot |x|^{-\alpha}, \quad \text{if } x \to 0, \alpha \geq d.$$

The proof is again founded on the checking of compactness conditions, but it uses the law of conservation of energy instead of the law of conservation of the number of particles.

We explain the main idea of the proof at a rather non-mathematical level. Let $r(t)$ be the radius of the ball with center μ such that the particle situated at μ at moment 0 will interact before moment t only with particles situated in this ball at moment 0. By the conditon 2) the general energy of the particles can not be more than $C \cdot [r(t)]^d$ (It is proportional to the volume of the ball). So velocities of each of particls can not be more than

$$v(t) \leq C_1^* \cdot [v(t)]^{d/2}.$$

If such an estimate for velocities is true for all particles we obtain that

$$\frac{dr(t)}{dt} \leq v(t) \leq C_1 \cdot [r(t)]^{d/2}.$$

Consider the differential equation

$$\frac{dx(t)}{dt} = C \cdot |x(t)|^{\gamma}.$$

When $\gamma \leq 1$, there is a continuous solution on the entire time axis; when $\gamma > 1$ its solution exists only in some finite interval and has a vertical asymptote. From this result and the comparison theorem we know that if $d = 1$ or 2, we can get a finite bound for the solution of the equation

$$\frac{dr(t)}{dt} \leq C \cdot [r(t)]^{d/2}, \quad \text{for all } t.$$

It can give us the necessary compactness estimate of a dynamical system. But if $d = 3$, we can not use this method to solve the problem and thus the problem is open.

We can really expect that for a good enough initial realization one particular particle may gain an essential part of the energy of other ones and so its velocity will go to infinity in a finite time. So I do not expect that in dimension $d = 3$ we can find a nice explicit condition on the initial realization which can guarantee the existence of the dynamics. I can only suppose that for some explicit conditions about the set of positions q_i of the particles at the initial moment and about the moduli $|v_i|$ of the velocities of the particles, the dynamics exists for almost all choices of the directions of velocities v_i. Such a type of results would be adequate for applications (See §4). It seems that the problem of the construction of infinite particle dynamics in dimension 3 is one of the most important, explicitly formulated open problems in mathematical statistical mechanics.

Now we consider the one-dimensional hard rod system to which we will return in the following sections. Because of its simplicity this system gives as a good laboratory for checking all the hypotheses. For the convenience of mathematical discussion, we regard the exchange of velocities between the two particles as an exchange of their "identities". After collision the particle 1 will be regarded as the particle 2 with the velocity of the original particle 1, and the same view is held for the particle 2 (See Fig(i)). The change will not be influenced by the parameters of the system.

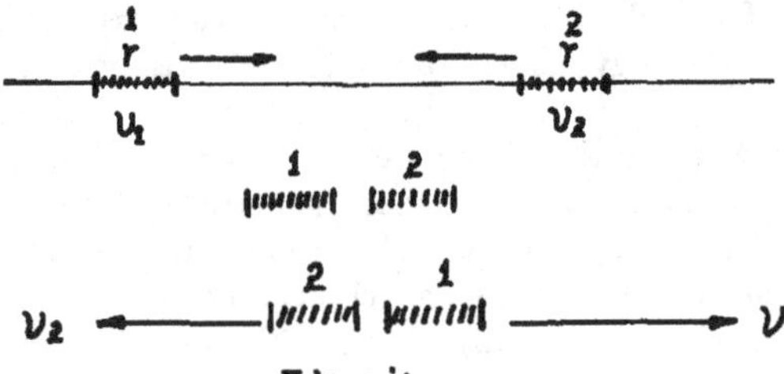

Fig (i)

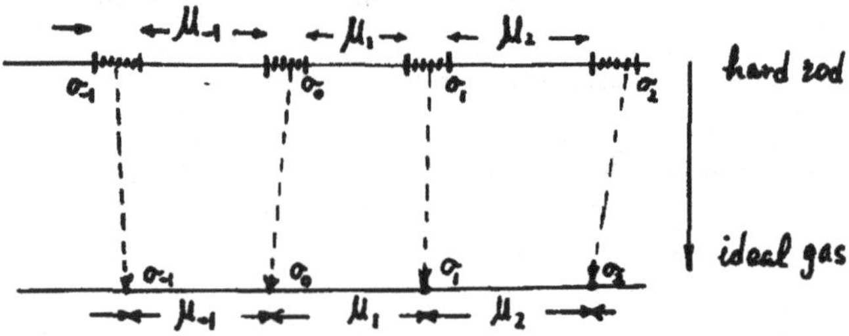

Fig(ii)

For the hard rod system modified above, we can use an ideal gas system as a representation of it. Figure (ii) explains the correspondence.

Here we have removed all hard rods and shifted the particles to a centrum particle conserving its velocities.

After studying the ideal gas model, we should return to the original model. To do this, first of all, we have to change the position at moment t of our particular particle which was a centrum at moment o. To this end we have to count the number $N_+(t)$ of other particles which intersect its trajectory before moment t in the ideal gas dynamics having velocities which are higher than the velocity of our particle. Similarly, we count the number $N_-(t)$ of particles with velocities lower than the velocity of our particle. Then we have to move the position of our particle at moment t for a distance $d \cdot (N_+(t) - N_-(t))$. At the same time, we move the positions of all other particles in a similar way. At last we have to make again a transformation opposite to the transformation of Fig. (ii).

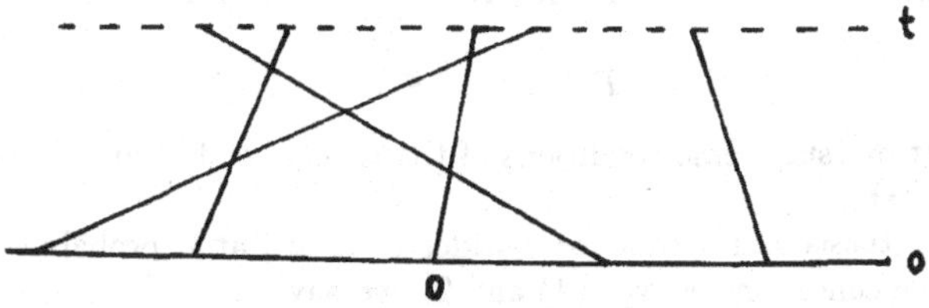

(Here $N_+(t) = 2, N_-(t) = 1$). This construction enables us to prove the existence of infinite particle dynamics for the model.(See [Dobrushin, Suhov (1979)] for details.)

§4. Random Evolution

In the previous sections, we discussed the non-random dynamics of infinite particle dynamical systems. Now we begin to study the random evolution which arises if the initial configuration is random. Let

$$T_t : a \to a(t), \text{for } a \in \mathcal{N} \subset \Omega \text{ and } t \in \mathbb{R}^1$$

be the dynamics described in the previous section and P is a probability measure on $\mathcal{A}$. We define P_t by setting

$$P_t(A) = P(T_{-t}A), \quad A \subset \mathcal{N}, A \in \mathcal{B}_\Omega, t \in \mathbb{R}^1.$$

We will call the family $\{P_t : t \in \mathbb{R}^1\}$ as an evolution of the initial state $P_0 = P$.

Very often we will need some restrictions on the class of possible initial measures P. There are two classes of conditions which seem useful and natural from the physical point of view. We will formulate them only in general terms because in different concrete problems different variants of their mathematical formulation can naturally arise.

1) Smoothness.

For any bounded open set $V \subset \mathbb{R}^d$, the restriction P_V of P on $\{a_V : a \in \mathcal{N}\}$ has a density p_V with respect to Lebesgue measure λ_V (See §1), i.e. $p_V = dP_V/d\lambda_V$. We can also introduce some conditons of smoothness about p_V, or of finiteness of the moments of some functionals of a_V, defined by the density p_V and so on.

2) Decay of correlation on large distances.

For example, for any two bounded open V_1 and V_2 of $\mathbb{R}^d$, define

$$d(V_1, V_2) = \sup_{\substack{A_1, A_2 \\ A_i \subset \mathcal{N}, A_i \in B_{\Omega_{V_i}} \\ i=1,2}} |P(A_1 \cap A_2) - P(A_1)P(A_2)|.$$

We can suppose that if $\mathrm{dist}(V_1, V_2) \to \infty$, then $d(V_1, V_2) \to 0$ exponentially fast. We can also use some other variants of conditions of mixing used in the probability theory.

As usual we will call a state (=probability measure) P invariant for a dynamics if $P(\mathcal{N}) = 1$ and

$$P_t \equiv P \quad \text{for all} \quad t \in \mathbb{R}^1.$$

If for this invariant measure some conditions of the type 1) and 2) are true we will call P an equilibrium state.

A natural hypothesis states that for a wider class of initial probability measures $\{P\}$ satisfying some conditions of types 1) and 2), we have

$$T_t P = P_t \Longrightarrow \bar{P}_0 \quad \text{as} \quad t \to \infty,$$

where $\bar{P}_0$ is an equilibrium state. We will always treat convergence as a convergence in the weak sense with respect to the topology of Ω introduced in §3.

Now we consider the three laws of conservation in their probabilistic interpretation. For any compact subset $W \subset \mathbb{R}^d, a_W = \{(q, \nu) : q \in W\}$, we let $N(a_W) = \sum_{(q,\nu) \in a_W} 1$, and consider the limit

$$\lim_{W \to \mathbb{R}^d} \frac{N(a_W)}{|W|} = N(a),$$

where $W \to \mathbb{R}^d$ in Van-Hove sense (See Sect. 3)

The law of conservation of particles is

$$N(a) = N(T_t a), t \in \mathbb{R}^1.$$

Let

$$\bar{N}_p = < N(a) >_P = \int N(a) dP.$$

We say that for an initial state P the system has the law of conservation of particles in the probability sense if $\bar{N}_{P_t} = \bar{N}_P$ for all $t \in \mathbb{R}^1$. Of course we must assume that the probability P is concentrated on the set in which the limit $N(a)$ exists.

We can also define other probabilistic laws of conservation by using the following notations.

We let

$$M(a_W) = \sum_{(q,\nu)\in a_W} \nu,$$

$$M(a) = \lim_{W\to\mathbb{R}^d} \frac{V(a_W)}{|W|},$$

$$\bar{M}_p = \langle V(a)\rangle_P,$$

and will say that the probabilistic law of conservation of moment is true if $\bar{V}_{P_t} \equiv \bar{V}_P$ for all t.

For energy, we let

$$E(a_W) = \sum_{(q,\nu)\in a_W} \frac{|\nu|^2}{2} + \sum_{(q_1,\nu_1),(q_2,\nu_2)\in a_W} U(|q_1 - q_2|),$$

$$E(a) = \lim_{W\to\mathbb{R}^d} \frac{E(a_W)}{|W|},$$

$$\bar{E}_P = \langle E(a)\rangle_P,$$

and will speak about the probabilistic law of conservation of energy if $\bar{E}_{P_t} \equiv \bar{E}_P$ for all t.

We expect that the support of equilibrium state P is so large that all three laws of conservation hold in the probabilistic sense, that is,

$$(\bar{N}_P, \bar{V}_P, \bar{E}_P) = (\bar{N}_{P_t}, \bar{V}_{P_t}, \bar{E}_{P_t}).$$

It is also natural to expect that if for an initial state P its evolution P_t converges to some equilibrium state $\overline{P}$, then

$$(\bar{N}_P, \bar{V}_P, \bar{E}_P) = (\bar{N}_{\overline{P}}, \bar{V}_{\overline{P}}, \bar{E}_{\overline{P}}).$$

Because these parameters can be chosen in an arbitrary way we expect the existence of at least a $(d+2)$-dimensional family of equilibrium states (the law of conservation of moment gives us d conserved parameters). Such a family really exists. We discuss its construction in the next section.

§5. Gibbsian States in Finite Volumes

We will begin the discussion with the case of motion in a compact volume W. As we explained in Section 2 we have here a well-defined dynamics under elastic boundary conditions, having laws of conservation of the number of particles and of energy. But we will consider now the case of periodic boundary conditions where we also have a law of conservation of moment.

We fix some constants $\bar{N}, \bar{V}, \bar{E}$ and consider a submanifold of W such that $N(a_W) = [\bar{N}\cdot|W|], E(a_W) = \bar{E}\cdot|W|, M(a_W) = \bar{M}\cdot|W|$, where $[\alpha]$ is the integer part of $\alpha \geq 0$. This submanifold is denoted by $\mathcal{M}_W(\bar{N}, \bar{V}, \bar{E})$.

Because of the laws of conservation this submanifold is invariant with respect to the dynamics. Under some natural conditions about the potential it is easy to check that this submanifold will be a smooth one. So it is possible to introduce the Lebesgue measure on it, and this measure will be invariant with respect to the dynamics. In the general case this measure is finite, and by normalizing it we obtain an invariant probability measure on $\mathcal{M}_W(\bar{N}, \bar{V}, \bar{E})$ which is called the microcanonical Gibbs state (or microcanonical ensemble). If we consider the non-periodic boundary condition for which the law of conservation of momentum is not true it is possible to fix only $\bar{N}$ and $\bar{E}$. In such a situation we also speak about the microcanonical Gibbs state. If we will fix only $\bar{N}$ we will call the corresponding invariant measure the small canonical Gibbs state (or small canonical ensemble).

Of course any convex linear combination of microcanonical Gibbs states with different values of $\bar{N}, \bar{V}, \bar{E}$ gives us again an invariant probability measure. So we can construct a new family of invariant probability measures on the set of configurations on Ω_W which are defined by

$$P_\alpha(a_W) = \frac{1}{Z_{W,\alpha}} \exp\{-[\alpha_N N(a_W) + \alpha_M \cdot M(a_W) + \alpha_E E(a_W]\},$$

where $\alpha_N, \alpha_E \in \mathbb{R}^1$ and $\alpha_M \in \mathbb{R}^d; \alpha_M \cdot M(a_W)$ is the scalar product in $\mathbb{R}^d$ and the normalizing constant is

$$Z_{W,\alpha} = \int \exp\{-[\alpha_N N(a_W) + \alpha_M \cdot M(a_W) + \alpha_E E(a_W)]\}\lambda_W(da).$$

This constant is called a partition function. It is easy to prove in many situations that $Z_{W,\alpha} < \infty$. We call this measure P_α the Gibbsian measure (or grand canonical Gibbs state) on Ω_W. Of course, P_α depends on $\alpha = (\alpha_N, \alpha_M, \alpha_E)$.

The foundation for such a definition will become clear later when we consider an infinite particle state in $\Omega_{\mathbb{R}^d}$.

For fixed values $\bar{N}_0, \bar{V}_0, \bar{E}_0$ we can choose parameters $\alpha = (\alpha_N, \alpha_V, \alpha_E)$ in such a way that the mean values $\langle\cdot\rangle_\alpha$ with respect to the measure P_α satisfy

$$\langle N(a_W)\rangle_\alpha = \bar{N}_0|W|,$$
$$\langle M(a_W)\rangle_\alpha = \bar{M}_0|W|,$$
$$\langle E(a_W)\rangle_\alpha = \bar{E}_0|W|.$$

It is possible to do so because these mean values are monotone functions of corresponding parameters (The domain of the change of $\bar{E}_0$ is bounded from below by a constant, depending on the potential). The following computation shows the monotonicity of parameter α_N.

$$E_\alpha N = \int N(a_W)\frac{1}{Z_{W,\alpha}} \exp\{-[\alpha_N N(a_W) + \alpha_M \cdot M(a_W) + \alpha_E E(a_W)]\}d\lambda_W.$$

So

$$\begin{aligned}
&\frac{\partial}{\partial \alpha_N} \ln Z_{W,\alpha} \\
=& -\frac{1}{Z_{W,\alpha}} \int N(a_W) \exp\{-[\alpha_N N(a_W) + \alpha_M \cdot M(a_W) + \alpha_E E(a_W)]\} d\lambda_W \\
=& -E_\alpha N
\end{aligned}$$

and

$$\frac{\partial}{\partial \alpha_N} E_\alpha N = -\frac{\partial^2}{\partial \alpha_N^2} \ln Z_{W,\alpha} = Var(N(a_W)) > 0,$$

where $Var(N(aW))$ is the variance of $N(a_W)$. The computation for other parameters is similar.

If the law of large numbers is true when $W \to \infty$, we will see that in the limit when $W \to \infty$, the canonical Gibbsian measure is asymptotically coincident with the microcanonical one for correspondingly chosen values of the parameter. However, this result is now proved only for the case where $\bar{N}_0$ or $\bar{E}_0$ is sufficiently small. It is impossible to expect this for large values of $\bar{N}_0$ and $\bar{E}_0$ because of the possibility of phase transition. We will not touch here the difficult problems of phase transition and will consider only the case of sufficiently small $\bar{N}_0$ and $\bar{E}_0$.

Physicists like to write the above measure in another way:

$$\frac{1}{Z_W} \exp\{-\beta[-\mu N(a_W) - \hat{\nu} V(a_W) + E(a_W)]\}.$$

The coefficients $\beta, \mu, \hat{\nu}$ are connected with the coefficients $\alpha_N, \alpha_V, \alpha_E$ in a one-to-one way. The physical meaning of the coefficients is as follows: $\beta = 1/T$, where T is the temperature of the system. Parameter μ is called a chemical potential. If $\mu \to \infty (\mu \to -\infty)$, then $\rho \to \infty (\rho \to 0)$, where $\rho = EN(a_W)/|W|$ is the density. So this parameter characterizes the density of the system. Since

$$E(a_W) = \frac{1}{2} \sum_i |\nu_i|^2 + \sum_{i \neq j} U(|q_i - q_j|),$$

and

$$\beta(\frac{1}{2} \sum_i |\nu_i|^2 - \sum_i \hat{\nu} \cdot \nu_i) = \beta(\frac{1}{2} \sum_i (\nu_i - \hat{\nu})^2 + const.),$$

then for any $a_W \in \Omega_W$, we have

$$\begin{aligned}
&P(a_W) \\
=& \frac{1}{Z_W} \exp\{-\beta[-\mu N(a_W) - \hat{\nu} V(a_W) + E(a_W)]\} \\
=& \frac{1}{Z_W} \exp\{-\beta(\frac{1}{2} \sum_i (\nu_i - \hat{\nu})^2 + const.) - \beta(-\mu N(q_W) \\
&- \sum_{q_1, q_2 \in q_W} U(|q_1 - q_2|))\} \\
=& \prod_i (Z')^{-1} \exp\{-\frac{\beta}{2} \cdot (\nu_i - \hat{\nu})^2\} \cdot P(q_W)
\end{aligned}$$

where

$$Z' = \int \exp\{-\frac{\beta}{2}(\nu - \hat{\nu})^2\} d\nu$$

(The integral is taken with respect to the usual Lebesgue measure on $\mathbf{R}^d$) and

$$P(q_W) = \frac{1}{Z''_W} \exp\{-\beta[-\mu N(q_W) + \sum_{q_1, q_2 \in q_W} U(|q_1 - q_2|)]\},$$

$$Z''_W = \int \exp\{-\beta[-\mu N(q_W) + \sum_{q_1, q_2 \in q_W} U(|q_1 - q_2|)\} \lambda''_W(dq_W),$$

$$q_W = \{q_i : (q_i, \nu_i) \in a_W\}, N(q_W) = N(a_W),$$

and λ''_W is the projection of the measure λ_W on the space of configurations.

It is easy to see that

$$\frac{1}{Z'} \exp\{-\frac{\beta}{2}(\nu - \hat{\nu})^2\}$$

is a Gaussian distribution. In physics one calls it the Maxwell distribution. Clearly, the velocities of particles are independent of one anothers and are independent of the position of particles (under the condition that the number of particles is fixed). The velocity of each particle is a random variable with Gaussian distribution with mean value $\hat{\nu}$ and variation β^{-1}.

Because the probabilistic structure of the velocities is very simple it is natural to concentrate on the study of probability densities $P(q_W)$. We will call them the configurational Gibbsian density.

Sometimes we will consider a configurational Gibbsian density corresponding to the dynamics in W with a boundary potential $\bar{U}(q)$ (See Sect.2), where $\bar{U}(q)$ has a finite range, i.e. $\bar{U}(q) \equiv 0$ if $|q| > r$. It has a density

$$P_W(q_W) = \frac{1}{Z_W} \exp\{-\beta[-N(q_W) + \sum_{\substack{q_1, q_2 \in q_W \\ q_1 \neq q_2}} U(|q_1 - q_2|) + \sum_{q \in q_W} \bar{U}(q)]\},$$

$$q_W \subset W,$$

where

$$Z_W = \int \exp\{-\beta[-N(q_W) + \sum_{\substack{q_1, q_2 \in q_W \\ q_1 \neq q_2}} U(|q_1 - q_2|) + \sum_{q \in q_W} \bar{U}(q)]\} d\lambda_W.$$

Now we discuss the Markov property of the density P_W. We suppose that $U(q) = 0$ if $|q| > M$. Let $\tilde{W} \subset W$ satisfy $\text{dist}(\tilde{W}, W^c) \geq r$ and $\partial_r \tilde{W} = \{x \in W \setminus \tilde{W} : dist(x, \tilde{W}) \leq r\}$. In analogy with the usual definition of Markov processes we can treat $\tilde{W}, \partial_r \tilde{W}$ and $(\tilde{W} \cup \partial_r \tilde{V})^c$ as "the future", "the present" and "the past" respectively.

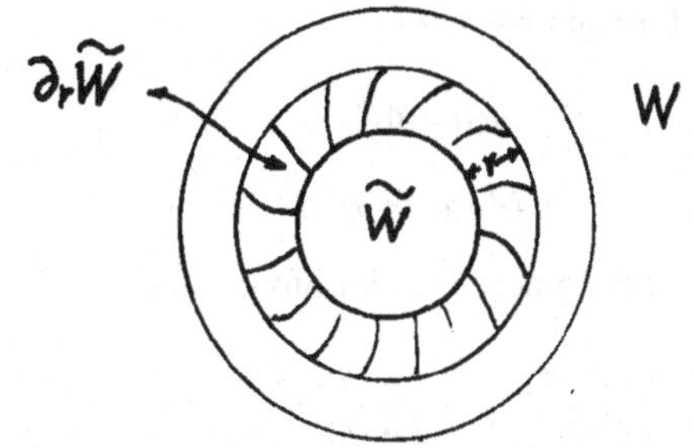

As we did in Section 2, we can introduce a basic measure on the configurations without considering the velocities. This measure is again denoted by λ_W. Then we have

$$\lambda_W = \lambda_{W\setminus\tilde{W}} \times \lambda_{\tilde{W}}.$$

Therefore the marginal density of the restriction of q_W on $W \setminus \tilde{W}$ is

$$P_W^{W\setminus\tilde{W}}(q_{W\setminus\tilde{W}}) = \int P_W(q'_{\tilde{W}} \cup q_{W\setminus\tilde{W}})\lambda_{\tilde{W}}(dq'_{\tilde{W}}).$$

And so we have the conditional density:

$$P_W(q_{\tilde{W}}|q_{W\setminus\tilde{W}}) = \frac{P_W(q_{\tilde{W}} \cup q_{W\setminus\tilde{W}})}{P_W^{W\setminus\tilde{W}}(q_{W\setminus\tilde{W}})}.$$

We will speak about the r-Markov property if for any $\tilde{W} \subset W$, we have

$$P_W(q_{\tilde{W}}|q_{W\setminus\tilde{W}}) = P_W(q_{\tilde{W}}|q_{\partial_r\tilde{W}}).$$

We claim that for the Gibbsian field this Markov property is true and

$$P_W(q_{\tilde{W}}|q_{\partial_r\tilde{W}}) \\ = \frac{1}{Z_W(q_{\partial_r\tilde{W}})} \exp\{-\beta[-\mu|q_{\tilde{W}}| + E(q_{\tilde{W}}|q_{\partial_r\tilde{W}})]\},$$

where the conditional energy is

$$E(q_{\tilde{W}}|q_{\partial_r\tilde{W}}) = \sum_{q_1,q_2\in q_{\tilde{W}}} U(|q_1 - q_2|) \\ + \sum_{q_1\in q_{\tilde{W}},q_2\in q_{W\setminus\tilde{W}}} U(|q_1 - q_2|) + \sum_{q\in q_{\tilde{W}}} \bar{U}(q),$$

and the conditional partition function becomes

$$Z_W(q_{\partial_r \tilde{W}}) = \int \exp\{-\beta[-\mu|q_{\tilde{W}}| + E(q_{\tilde{W}}|q_{\partial_r \tilde{W}})]\}\lambda_{\tilde{W}}(dq_{\tilde{W}}).$$

To prove this fact it is enough to note that for any $q_{\tilde{W}}, q'_{\tilde{W}}, q_{W\setminus\tilde{W}}$, we have

$$\begin{aligned}\frac{P_W(q_{\tilde{W}}|q_{W\setminus\tilde{W}})}{P_W(q'_{\tilde{W}}|q_{W\setminus\tilde{W}})} &= \frac{P_W(q_{\tilde{W}}\cup q_{W\setminus\tilde{W}})}{P_W(q'_{\tilde{W}}\cup q_{W\setminus\tilde{W}})}\\ &= \exp\{-\beta[-\mu(|q_{\tilde{W}}|-|q'_{\tilde{W}}|)\\ &\quad + E(q_{\tilde{W}}|q_{W\setminus\tilde{W}}) - E(q'_{\tilde{W}}|q_{W\setminus\tilde{W}})]\}\\ &= \frac{P_W(q_W|q_{\partial_r \tilde{W}})}{P_W(q'_W|q_{\partial_r \tilde{W}})}.\end{aligned}$$

§6. Gibbsian Measures in an Infinite Volume

In the previous section we defined Gibbsian measures in finite volumes. Now we want to define Gibbsian measures on the whole configuration space $\Omega = \{\{q,\nu)\} : q,\nu \in \mathbb{R}^d\}$ (See §1). Recall that for $a = \{(q,\nu)\} \in \Omega$ and a bounded subset $W \subset \mathbb{R}^d$, we have

$$a_W = \{(q,\nu) \in a : q \in W\},$$
$$\Omega_W = \{a_W : a \in \Omega\},$$

and for each $\tilde{W} \subset W$,

$$\Omega_W = \Omega_{\tilde{W}} \times \Omega_{W\setminus\tilde{W}},$$
$$\lambda_W = \lambda_{\tilde{W}} \times \lambda_{W\setminus\tilde{W}},$$

where λ_W is the basic measure on Ω_W constructed in Section 1.

For simplicity, we consider only the potential U with a finite range (i.e. $U(|q|) = 0$ if $|q| > r$), and now we suppose that there is no external field (i.e. $\bar{U}(q) \equiv 0$). We first define the configurational Gibbsian measures for positions of the particles. Let $\mathcal{A}^1 = \{q = \{q_i\} : q_i \in \mathbb{R}^d\}$. Let $\mathcal{B}^1$ be the corresponding σ-algebra of subsets (see section 1.) For each $q \in \mathcal{A}^1$ and $W \subset \mathbb{R}^d$, let

$$q_W = \{q_i \in q : q_i \in W\},$$
$$\mathcal{A}^1_W = \{q_W : q \in \mathcal{A}^1\}$$

and

$$\partial W = \{x \in W^c : dist(x, W) \le r\}.$$

For any bounded $W \subset \mathbb{R}^d$ and $\bar{q}_{W^c} \in \mathcal{A}^1_{W^c}$, define

$$P_W(q_W|\bar{q}_{W^c}) = \frac{1}{Z(\bar{q}_{W^c})}\exp\{-\beta[-\mu N(q_W) + H_W(q_W|\bar{q}_{W^c})]\}$$

where

$$H_W(q_W|q_{W^c}) = \sum_{\substack{q_1,q_2\in q_W\\ q_1\neq q_2}} U(|q_1-q_2|) + \sum_{\substack{q_1\in q_W\\ q_2\in q_{W^c}}} U(|q_1-q_2|),$$

$$Z(\bar{q}_{W^c}) = \int \exp\{-\beta[-\mu|q_W| + H_W(q_W|\bar{q}_{W^c})]\}d\lambda^1_W,$$

where λ^1_W is the marginal measure of λ_W.

We call $P_W(q_W|\bar{q}_{W^c})$ a conditional Gibbsian density in a finite volume W with boundary condition $\bar{q}_{W^c}$. Clearly, $P_W(q_W|\bar{q}_{W^c})$ depends only on a restriction $\bar{q}_{\partial W} = \bar{q}_W|_{\partial W}$. So we can write it as $P_W(q_W|\bar{q}_{\partial W})$. Of course this definition is suggested by the corresponding relation in a finite volume discussed in the previous section.

In order to define Gibbsian measure on $\mathcal{A}^1$, we need to construct a σ-algebra for each $W \subset \mathbb{R}^d$ which describes the behaviour of the restriction of the field to W. We define $\mathcal{B}^1_W$ as the smallest σ-algebra with respect to which all functions $f(X) \triangleq |X \cap S|, X \in \mathcal{A}^1_W$ are measurable, where $S \subset W$ is any bounded domain.

Given a probability measure P on $(\mathcal{A}^1, \mathcal{B}^1)$ let $P(\cdot|\mathcal{B}'_{W^c})$ be the conditional probability measure of P on $(\mathcal{A}^1_W, \mathcal{B}^1_W)$ under σ-algebra $\mathcal{A}^1_{W^c}$. For each $\bar{q}_{W^c} \in \mathcal{A}^1_{W^c}$, let $P(\cdot|\bar{q}_{W^c}) = P(\cdot|\mathcal{B}^1_{W^c})(\bar{q}_{W^c})$.

Now we can give the definition of Gibbsian measure.

Definition. A probability measure P on $(\mathcal{A}^1, \mathcal{B}^1)$ is said to be a configurational Gibbsian measure if for any bounded subset $W \subset \mathbb{R}^d, P(\cdot|\bar{q}_{W^c})$ has a density $P_W(q_W|\bar{q}_{\partial W})$ with respect to λ^1_W for P-almost every $\bar{q}_{W^c} \in \mathcal{A}^1_{W^c}$.

We have defined the positional Gibbsian measure which depends on the two parameters μ and β. Then we also have to define the Gibbsian measure for position and velocity, which depends on $d+2$ parameters $(\mu, \beta, \hat{\nu})$. In the finite volume case the velocities have Gaussian distributions with densities

$$P_{\hat{\nu}}(\nu) = (\frac{\beta}{2\pi})^{-d/2}\exp\{-\frac{\beta}{2}\sum_i(\nu_i - \hat{\nu}_i)^2\}.$$

Morevoer, the velocities of different particles are independent under the condition that the positions are fixed. So it is natural to define the following conditional Gibbsian density for position and velocity:

$$P_W(a_W|\bar{a}_{\partial W}) = \prod_{(q,\nu)\in a_W} P_{\hat{\nu}}(\nu)\cdot P_W(q_W|\bar{q}_{\partial W}),$$

where $a_W = \{(q,\nu) : q \in W\} \in \Omega_W, \bar{a}_{\partial W} \in \Omega_{\partial W}$. Then we can introduce a definition of Gibbsian measure on Ω in a complete analogy with the previous definition.

Now an immediate question arises: Does a Gibbsian measure exists, and if so, when is it unique? The existence of Gibbsian measures can be proved under very general circumstances. Furthermore, if parameter β is sufficiently small and μ sufficiently close to - ∞ then there exists only one Gibbsian measure; in the other cases, it is natural to expect that the Gibbsian measure is not unique even though there are no mathematically proven examples of the non-uniqueness for such a situation. The non-uniqueness results are known mainly for lattice models (See Sinai (1982).) In the case where it is possible to prove the uniqueness of a Gibbsian field, it is also possible to prove that the conditions of decay of the correlations discussed in §4 are fulfilled.

The general theorem of functional analysis implies that the set of Gibbsian measures with a certain potential is a convex set. If we choose an appropriate sequence of subsets $\{W_n : n \geq 1\}$, $W_n \subset W_{n+1} \subset \mathbb{R}^d$ and an appropriate boundary condition $\bar{a}_{\mathbb{R}^d \setminus W_n}$, we can construct any extremal Gibbsian state as

$$\lim_{n\to\infty} P_{W_n}(\cdot|\bar{a}_{\mathbb{R}^d\setminus W_n}) = P(\cdot).$$

The convergence is a weak convergence with respect to the topology on Ω defined previously.

§7. Random Evolution (revisited)

Having the notion of Gibbsian state in an infinite volume we can return to the questions discussed in §4.

First of all we can state more accurately what type of results concerning the existence of dynamics we want to have. We expect that the dynamics exists for almost all initial realizations with respect to a wider class of Gibbsian states with potentials, which define the dynamics. For the case of dimension $d = 1$ or $d = 2$, this follows from the results described in §3. For the case $d \geq 3$, the problem is still open.

In a paper by Presutti, Pulverenti and Tirozzi (1976), it was proven that for a wide class of potentials of dynamics, the dynamics exists for almost all initial distributions having the same potential as that of the dynamics, and all Gibbsian fields with this potential are invariant for the dynamics. This result can be treated as a limit variant of the invariance in finite volume discussed in §4 and §5.

To justify completely the identification of the classes of Gibbsian and equilibrium measures, as it is usual in physical literature, we also have to show that any invariant measure satisfying conditions of §4 is a Gibbsian measure with a potential coinciding with the potential of the dynamics. Important results in this direction are due to Gurevich and Suhov (1976-1982). To explain their results we have to introduce another description of evolution, generally used in physics literatures.

For any $A \subset (\mathbb{R}^d \times \mathbb{R}^d)^n$, let

$$K^{(n)}(A) = E_p\{\sum_{\substack{a'\subset a\\ |a'|=n}} \chi_A(a')\}_P,$$

where P is a probability measure on Ω, χ_A is the indicator of set A. It is easy to check that $K^{(n)}(A)$ is a measure on $(\mathbb{R}^d \times \mathbb{R}^d)^n$ (generally not a probability measure). If $K^{(n)}(\cdot)$ has a density $k_n(\cdot)$ with respect to $\lambda_{\mathbb{R}^d}$, following a tradition in physics, we call $\{k_n(\cdot) : n \geq 1\}$ a correlation function (In this place the terminology of physics is not consistent with that of probability theory, where the correlation function has another meaning).

For any $a' = \{(q_1, \nu_1), \cdots, (q_n, \nu_n)\}$, when $\varepsilon > 0$ is small enough, we have

$$P(\cap_{i=1}^{n}\{a : |a_{B(q_i,\nu_i,\varepsilon)}| = 1\})$$
$$\approx k_n(a')\prod_{i=1}^{n}|B(q_i, \nu_i, \varepsilon)|,$$

where $a_{B(q_i,\nu_i,\varepsilon)}$ is the restriction of a on the ball $B(q_i, \nu_i, \varepsilon)$ with center $x_i = (q_i, \nu_i)$ and radius ε. This explains the probabilistic sense of this definition.

In general, if for any $W \subset \mathbb{R}^d, |W| < \infty$, we have

$$\sum_n \frac{1}{n!}K^{(n)}(W \times \mathbb{R}^d) < \infty,$$

then $\{k_n(\cdot) : n \geq 1\}$ are uniquely defined the measure P. This property can be checked for a wide class of Gibbsian states.

It is natural to expect that the correlation functions $\{k_n(t) : t \geq 0\}_{n\geq 1}$ of the family of states $\{P_t, t \in \mathbb{R}_+\}$ describing the evolution of the dynamics satisfy the well-known BBGKY hierarchy. The BBGKY equations are as follows:

$$\frac{\partial}{\partial t}k_n(a,t) = \{k_n(a, H(a))\}$$
$$+ \int dq_0 d\nu_0 k_{n+1}((a, (q_0, \nu_0)), \sum_{j=1}^{n} U(|q_0 - q_j|))$$

where $\{\cdot\}$ is the Poisson parents.

In the case of finite volume it is easy to check that BBGKY - equations are essentially equivalent to the hypothesis that the corresponding family of states is obtained by the help of the evolution with potential U. For the general case, similar problem has not been thoroughly investigated (See Gallavott, Lanford, Lebowitz (1972), Sinai, Suhov (1974)).

The BBGKY equations are difficult to be investigated, because $\frac{\partial}{\partial t}k_n$ depends on not only k_n but also k_{n+1}. Gurevich and Suhov have investigated the question about the solutions of BBGKY equations which don't depend on t. They strictly restricted a priori the class of states $P_t \equiv P$ which they treated. It is a class of Gibbsian states with non- pair potentials. These states can be defined in the following way.

Let us consider an additive translation-invariant functional

$$H(a) = \sum_{(q,\nu)\in a}\phi_1(q,\nu) + \sum_{(q_1,\nu_1),(q_2,\nu_2)\in a}\phi_2(q_1 - q_2, \nu_1, \nu_2)+$$
$$+ \sum_{(q_i,\nu_i)\in a, i=1,2,3}\phi_3(q_1 - q_2, q_1 - q_3, \nu_1, \nu_2, \nu_3) + \cdots,$$

defined for finite realizations by a family of functions $\phi = \{\phi_1, \phi_2, \cdots\}$. Then we can define the Hamiltonian with any boundary condition as follows: For any bounded $W \subset \mathbb{R}^d$ let

$$H(a_W|a_{\mathbb{R}^d\setminus W}) = H(a) - H(a_{\mathbb{R}^d\setminus W}).$$

This definition can be extended by a natural limit approach to infinite realizations. We can also define Gibbs state on Ω by the conditional densities $\{P_W(a_W|a_{\mathbb{R}^d\setminus W})\}$, where

$$P_W(a_W|a_{\mathbb{R}^d\setminus W}) = \frac{1}{Z}\exp\{-H(a_W|a_{\mathbb{R}^d\setminus W})\}$$

If ϕ vanishes when $|q_1 - q_2| \geq r$ for some i and some $r > 0$, then the Gibbs state with respect to ϕ has an r-Markov property. Of course, this definition includes the case considered in §6 as a special case when $\phi_k \equiv 0, k \neq 2$.

Now we can describe the Gurevich-Suhov result. They considered the class of Gibbsian states given by a potential ϕ under some strong additional conditions on ϕ. One of the conditions is that $\phi_k \equiv 0$ if $k > k_0$ for some k_0. They supposed that the system of correlation functions corresponding to this Gibbsian state satisfies the BBGKY-equations, and proved that it implies that the potential coincides with the potential of dynamics. It is known, that under some strong condition about its conditional probabilities any field has to be Gibbsian with some potential. So the Gurevich-Suhov result means that under some very strong variants of conditions of type 1) and 2) of Section 4, each equilibrium state is really a Gibbsian state.

A question about the convergence in the evolution of a wider class of initial states to equilibrum ones has been solved only for some very simple and degenerate models. The first is the ideal gas model, i.e. the case where the dynamical potential $U(x) = 0, |x| > 0$. In this model each particle moves along a straight line with constant velocity and without interacting with others. Because $U(q) \equiv 0$ in this case, the Gibbsian measure is a Poisson field with density λ depending on the chemical potential. Again we have equally distributed and mutally independent velocities under the condition that the positions are fixed. But now the probability distribution of the velocities can still be arbitrary because of the additional law of conservation of velocities in this model. We will suppose that the distribution is given by a density $f(\nu)$. Then the first correlation function of the equilibrium field is

$$r_1(q,\nu) = \lambda f(\nu),$$

and all other correlation functions are

$$r_k(q_1,\nu_1,\cdots,q_k,\nu_k) = \prod_{i=1}^{k} r_1(q_i,\nu_i).$$

So the equilibrium fields can be described by their first correlation functions. The fact that this state is invariant with respect to the dynamics was in essence proved already in Doob's book [Doob (1953)].

For the ideal gas model it is easy to prove that for a large class of translation-invariant initial states P_0 the evolution P_t converges as $t \to \infty$ to an equilibrium state with the same first correlation function as that for the initial state (See [Dobrushin 1956] and [Dobrushin, Suhov (1985)]). We explain the idea of the proof.

Fix a finite interval of position (or sphere if $d > 1$) and denote it by I (See Figure below).

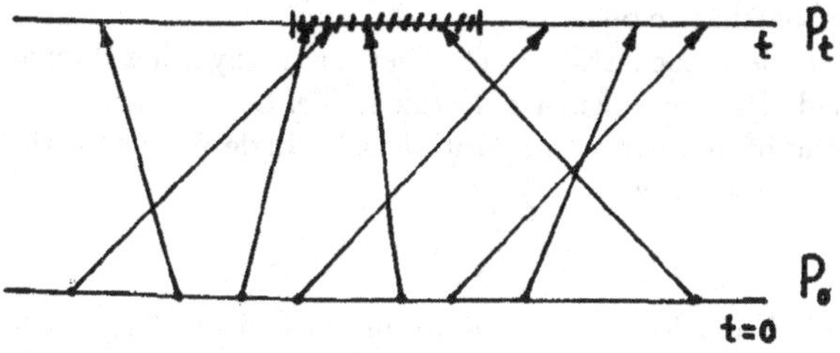

Because the velocity conserves, we can consider only how many particles with velocities in some small interval $\Delta\nu$ are in I at moment t. Because all the particles move with different velocities, hence the particles in I at time t which is large enough should come from quite different places. Because we suppose that the initial state P_0 has a property of decay of correlations, we can use a variant of Poisson limit theorem for weakly dependent variables to prove the convergence to an equilibrium measure. There is no other "sufficiently good" invariant measure for the dynamics.

The second model is the one-dimensional hard rods, i.e. the case where the dynamical potential $U(x) = \infty$ for $|x| \leq r$ and $U(x) = 0$ for $|x| > r$. A possibility of studying a system of one-dimensional hard rods is related to the possibility of reducing the system to that of an ideal gas, which was described in §2. The degeneracy of these models implies that velocity's probability distribution is again preserved during the evolution. Hence the model has invariant states corresponding to any such distribution. If this distribution has a density $f(\nu)$, then these invariant states may be described as Gibbs states with a Hamiltonian

$$H(a) = \sum_{(q_1,\nu_1),(q_2,\nu_2)\in a} U(|q_1 - q_2|) - \sum_{(q,\nu)\in a} (\ln f(\nu) - \mu).$$

This invariant state can again be described by its first correlation function. Here it is possible to prove [Dobrushin,Suhov (1985)] the results concerning the class of equilibrium measures and the convergence to them which are completely similar to the results explained above for the ideal gas model.

§8. Hydrodynamical Equations

Now we begin to discuss the so-called hydrodynamical limit approach. This problem was studied at a mathematical level firstly by Morrey (1955) who introduced a general definition of hydrodynamical limit approach. But in his paper he made a lot of hypotheses about the evolution of P_t for $t > 0$ which can not be proved to be right even now.

We illustrate the ideas by the following three models.

1) Real fluid dynamics.

For this model considered in previous sections we only give some definitions and formulate some hypotheses. There has been no mathematical results for this model.

2) One-dimensional hard rod model.

For this model it is possible to obtain complete enough results.

3) Brownian particles model.

This is the simplest stochastic model. There are many other stochastic models which have been studied. But we consider only this model here. In this model particles move independently one of another. The position of a single particle satisfies the following stochastic differential equation:

$$\frac{dx_i}{dt} = a + b\xi_i$$

where a is the shift and $b > 0$ is the diffusion coefficient. The processes ξ_i are independent for different i and $\xi_i(t)$ is the white noise. We know that the solution of this equation is a diffusion process and different particles are independent. In this system we have only one law of conservation-that of the number of particles. It is easy to prove that the Poisson point field is the unique equilibrium state of this dynamical system. In spite of the triviality of the model it is very convenient for illustrating all the main notions to be consider below.

In order to describe the hydrodynamical limit approach we need to introduce some new concepts. As we explained earlier if the initial state P_0 is translation invariant one can expect that the evolution P_t will converge to an equilibrium state. Now we will consider the non-translation invariant initial state. Of course it is impossible to say something interesting about the most general case. So we introduce the following two conditions on initial states.

Definition 1. We call a family of initial states $\{P_0^\varepsilon\}, \varepsilon \to 0$ locally translation-invariant if for any $q \in \mathbb{R}^d$,

$$(S_q P_0^\varepsilon - P_0^\varepsilon) \to 0, \quad \text{as } \varepsilon \to 0,$$

where $S_q P_0^\varepsilon(A) \triangleq P_0^\varepsilon(A - q), A \in \mathcal{B}_\Omega$ and the convergence is in the weak sense with respect to the topology of Ω introduced before.

Hydrodynamics studies the evolution of locally translation invariant states.

In a classical fluid the parameter ε characterizes a typical ratio of the space-time microscopic to macroscopic scales. It means that there is a small parameter ε such that ε^{-1} is much larger than the mean distance travelled by a particle during a unit time and that the initial state changes only a little under shifts over a distance which is much smaller than ε^{-1}. So from the microscopic point of view we can regard that P_0^ε is translation invariant if ε is small enough. When $t \to \infty$, $\varepsilon \to 0$ and $t << \varepsilon^{-1}$, we could expect that P_t^ε asymptotically approximate an equilibrium state. So the following definition is natural.

Definition 2. We call a family of initial states $\{P_0^\varepsilon\}$ a local equilibrium family if there is a family of equilibrium states $\{\hat{P}_{0,q}^\varepsilon\}, q \in \mathbb{R}^d$ such that

$$(S_q P_0^\varepsilon - \hat{P}_{0,q}^\varepsilon) \to 0, \quad \text{as } \varepsilon \to 0.$$

Now we give a brief description of meaning of evolution equations. Because a complete description of the states is too complex we want to study the evolution P_t of some system with initial state P_0 by studying a special, more simple functionals F of P_t. Set $F_t \triangleq F(P_t)$. The main difficulty is that generally F_t depends not only on F_0 but also on all the states P_0. So the transformation $F_0 \to F_t$ does not define a semigroup. By a scaling transformation, we get a family of states P_0^ε (respectively, F_0^ε). Then $F_0^\varepsilon \to F_t^\varepsilon$ if $P_0^\varepsilon \to P_t^\varepsilon$. Under some hypotheses on P_0^ε, F_0^ε, we could expect that the limit

$$\lim_{\varepsilon \to 0} F_t^\varepsilon = \bar{F}_t$$

exists in some sense. Then we have another evolution corresponding to the original one,

$$\bar{F}_0 \to \bar{F}_t \triangleq \bar{T}_t(\bar{F}_0).$$

And we have a family of $\{\bar{T}_t\}$ of transformations. We can hope that in some cases $\bar{F}_t$ depends only on $\bar{F}_0$ and so $\{\bar{T}_t\}$ is a semigroup:

$$\bar{T}_t \cdot \bar{T}_s = \bar{T}_{s+t}.$$

Then we can hope that $\bar{F}_t$ satisfies the following type of equations:

$$\frac{\partial \bar{F}_t}{\partial t} = A(\bar{F}_t).$$

Usuelly A is a nonlinear operator. It is possible to mention four types of such equations (See [Dobrushin, Sinai, Suhov (1985)] and the references there

1) Boltzmann equation
2) Vlasov equation
3) Landau equation
4) Hydrodynamical equation.

We will restrict ourselves here only to the consideration of the hydrodynamical limit approach.

Now we consider some different models.

1. Hard rods in one-dimensional case.

As we have explained above an equilibrium state here is defined by its first correlation function. We need to acquire some natural conditions about the initial states $\{P_0^\varepsilon\}$.

There exists a constant $\alpha > 0$ such that

$$\varphi_\varepsilon(d) \le e^{-\alpha d} \quad (\varepsilon > 0)$$

$$\varphi_\varepsilon(d) \triangleq \sup_{(V_1,V_2):dist(V_1,V_2)\ge d} \; \sup_{A\in\mathcal{B}(V_1) B\in\mathcal{B}(V_2)} |P_0^\varepsilon(A\cap B) - P_0^\varepsilon(A)P_0^\varepsilon(B)|.$$

Let us use $\bar{r}_0(q,\nu)$ to denote the first correlation function of initial state P_0. Define

$$r^\varepsilon(q,\nu,0) = \bar{r}_0(\varepsilon \cdot q, \nu).$$

This means that we transform the microscopic system to the microscopic system by a scaling of ε. Then we obtain a family of initial states P_0^ε with the given first order correlation functions $r^\varepsilon(q,\nu,0)$. Then we have the evolution P_t^ε with initial state P_0^ε. Let $r^\varepsilon(q,\nu,t)$ be a first correlation function of this state. Then we define

$$\bar{r}^\varepsilon(q,\nu,t) = r^\varepsilon(\varepsilon^{-1}\cdot q,\nu,\varepsilon^{-1}t).$$

This means that we again transform the micrscopic system back to the microscopic system by a scaling of ε^{-1}.

Theorem. Under the hypotheses which we gave previously we have the following results for the hard rod model. The limit

$$\bar{r}(q,\nu,t) = \lim_{\varepsilon\to 0} \bar{r}^\varepsilon(q,\nu,t)$$

exists for all t and we call $\bar{r}(q,\nu,t)$ the hydrodynamical function. The hydrodynamical function $\bar{r}(q,\nu,t)$ depends only on $\bar{r}_0(q,\nu)$. And the transformation $T_t : \bar{r}(q,\nu,0) \to \bar{r}(q,\nu,t)$ has a semigroup property, i.e., $T_s \cdot T_t = T_{s+t}$. Moreover, $\bar{r}(q,\nu,t)$ satisfies the following Euler type equation:

$$\frac{\partial \bar{r}(q,\nu,t)}{\partial t} = -\frac{\partial}{\partial q}[\bar{r}(q,\nu,t)$$
$$(\nu + \frac{d\int(\nu'-\nu)\bar{r}(q,\nu',t)d\nu'}{1-d\int \bar{r}(q,\nu',t)d\nu'})].$$

In the paper of Boldrighini, Dobrushin and Suhov (1983) the results have been proved.

Now we want to discuss how definitions there can be generalized to the case of real fluid dynamics. Because equilibrium distributions are defined by the parameters of the laws of conservation we have to follow not the first correlation functions but their densities. We have found that some laws of conservation play an important role in formulating the hydrodynamical equations. We use $\bar{n}_0(q), \bar{p}_0(q)$ and $\bar{e}_0(q)$ to denote the density of the number of particles, the densities of the momentum and of the energy respectively under the same equilibrium initial state P_0. Suppose that

1) $\int r_1^\varepsilon(q,\nu,0)d\nu = \bar{n}_0(\varepsilon q)$,

2) $(\int r_1^\varepsilon(q,\nu,0)\nu_i d\nu, i=1,2,3) = \overline{P}_0(\varepsilon q)$,

3) $\frac{1}{2}\int r_1^\varepsilon(q,\nu,0)|\nu|^2 d\nu + \frac{1}{2}\int r_2^\varepsilon(q,\nu_1,q_2,\nu_2,0)U(|\varepsilon q - q_2|)dq_2 d\nu_1 d\nu_2 = \bar{e}_0(\varepsilon q)$.

A quantity $\bar{n}_t(\cdot), \bar{p}_t(\cdot)$ and $\bar{e}_t(\cdot)$ can be defined in a similar way. It is expected that the usual Euler equations can be obtained for these quantities in a way similar to the described above.

2. Brownian particle model.

For some models, another variant of the hydrodynamics limit approach with another time normalization is useful. Instead of $\bar{r}_t(q) = \bar{r}(\varepsilon^{-1}t, \varepsilon^{-1}q)$, we take

$$\bar{\bar{r}}_t^\varepsilon(q) = r^\varepsilon(\varepsilon^{-2}t,\varepsilon^{-1}q).$$

It is easy to show that for Brownian particles if $a = 0$, then we have the following equation

$$\frac{\partial \bar{\bar{r}}_t(q)}{\partial t} = \frac{b^2}{2}\frac{\partial^2 \bar{\bar{r}}_t(q)}{\partial q^2}.$$

It is not true to call this equation Euler equation. If $a = 0$, we again can apply the normalization obtained by scaling time with $t \to \varepsilon^{-1}t$ and obtain the equation

$$\frac{\partial \bar{r}_t(q)}{\partial q} = 0.$$

So it is the case where the Euler equation is trivial.

Boldrighini, Dobrushin and Suhov (unpublished) consider another type of equation, that is, the equation of Navier-Stokes type. In the following, it is possible to suggest a construction for a equation of Navier-Stokes type. To be concrete, we shall describe the construction for the case of hard rod model and Brownian particle model. Generalization of the definitions to other models will be evident.

We suppose that the thermodynamical limit

$$\lim_{\varepsilon \to 0} \bar{r}^\varepsilon_t(q,\nu) = \bar{r}_t(q,\nu)$$

exists and that the Euler equation for $\bar{r}_t(q,\nu)$ is valid, where $\bar{r}^\varepsilon_t(q,\nu) = r^\varepsilon(\varepsilon^{-1}t, \varepsilon^{-1}q, \nu)$. We also assume that the limit

$$B\bar{r}_0(q,\nu) = \lim_{t\to 0}\lim_{\varepsilon\to 0}(\varepsilon t)^{-1}[\bar{r}^\varepsilon_t(q,\nu) - \bar{r}_t(q,\nu)]$$

exists (where B is in general a nonlinear operator and the above chosen order of limits is very essential). Then the equation

$$\frac{\partial \hat{r}_t(q,\nu)}{\partial t} = A\hat{r}^\varepsilon_t(q,\nu) + \varepsilon B\hat{r}^\varepsilon_t(q,\nu)$$

will be called an equation of Navier-Stokes type for the model considered, where A is the operator in the equation of Euler type.

For the hard rod model, we can prove that under some conditions

$$\lim_{t\to 0}\lim_{\varepsilon\to 0}(t\varepsilon)^{-1}(\bar{r}^\varepsilon_t(q,\nu) - \bar{r}_t(q,\nu))$$
$$= \frac{1}{2}\frac{\partial}{\partial q}(b(q,\nu)\frac{\partial \bar{r}_0(q,\nu)}{\partial q})$$

where

$$b(q,\nu) = \frac{d^2}{2}\frac{\partial}{\partial q}((1 - a\int \bar{r}_0(q,\nu)d\nu)^{-1}[\int |\nu - \nu^1|(q,\nu^1)\frac{\partial \bar{r}_0(q,0)}{\partial q}d\nu^1$$
$$-\bar{r}_0(q,0)\int d\nu^1|\nu - \nu^1|\frac{\partial}{\partial q}\bar{r}_0(q,0)]).$$

Therefore the corresponding equation of Navier-Stokes type is

$$\frac{\partial \hat{r}^\varepsilon_t(q,\nu)}{\partial t} = A\hat{r}^\varepsilon_t(q,\nu) + \frac{\varepsilon}{2}\frac{\partial}{\partial q}(b(q,\nu,t)\frac{\partial \hat{r}^\varepsilon_t(q,\nu)}{\partial q}),$$

where $b(q,\nu,t)$ is defined as above with $\bar{r}_0(q,v)$ changed on $\bar{r}_0^\varepsilon(q,v)$ and

$$A\hat{r}_t^\varepsilon(q,\nu) = -\frac{\partial}{\partial q}(\hat{r}_t^\varepsilon(q,\nu)[\nu + d\frac{\int(\nu'-\nu)\hat{r}_t^\varepsilon(q,\nu')d\nu'}{1-d\int\hat{r}_t^\varepsilon(q,\nu')d\nu'}]).$$

For the Brownian particle model, the corresponding Navier-Stokes equation is

$$\frac{\partial\hat{r}_t(q)}{\partial t} = -a\frac{\partial\hat{r}_t^\varepsilon(q)}{\partial q} + \frac{\varepsilon}{2}b^2\frac{\partial^2\hat{r}_t^\varepsilon(q)}{\partial q^2}.$$

This is easy to understand by an explicit computation.

One of the justifications of this definition consists in the expectation that the function $\hat{r}_t^\varepsilon$ which is a solution of the Navier-Stokes equation gives a better approximation to the true dynamics $\bar{r}_t^\varepsilon$ in comparision with the $\bar{r}_t$ which is a solution of Euler equation. If the limit

$$\lim_{\varepsilon\to 0}\bar{r}_t^\varepsilon(q,\nu) = \bar{r}_t(q,\nu)$$

exists, it is natural to expect only that

(1) $$\sup_{0\le t\le T}|\bar{r}_t^\varepsilon(q,\nu) - \bar{r}_t(q,\nu)| = o(1), \varepsilon\to 0 \text{ for any } T>0.$$

However, for the solution $\hat{r}_t^\varepsilon$ of the Navier- Stokes equation, one may hope that

(2) $$\sup_{0\le t\le T}|\bar{r}_t^\varepsilon(q,\nu) - \hat{r}_t^\varepsilon(q,\nu)| = 0(\varepsilon), \varepsilon\to 0 \text{ for any } T>0.$$

Furthermore, it is natural to expect that

(3) $$\sup_{0\le t\le \varepsilon^{-1}T}|\bar{r}_t^\varepsilon(q,\nu) - \hat{r}_t^\varepsilon(q,\nu)| = o(1), \varepsilon\to 0 \text{ for any } T>0.$$

These statements have not been proved yet even for the hard rod model. For Burwnian particl model they are trivial.

Added in proof (1992): I detailed discussion of the problems of this type on a simple model can be found in a recent paper [Dobrusnin, Sokolovskii (1991)].

References

Arnold V.I. (1978), Mathematical methods of Classical Mechanics. Springer-Verlag, New-York, 462p.

Boldrighini C., Dobrushin R.L., Suhov Yu.M. (1983), One-dimensional hard rod caricature of hydrodynamics. J.Stat.Phys., **31**, No.3, 577-615.

Chulaevsky F. A. (1983), The inverse problem method of scattering theory in statistical physics. Funct. Anal. and its Appl., **17**, No 1, 53-62.

Dobrushin R. L. (1956), On Poisson's law for distribution of particles in space. Ukrain. Math. J., 8, No 2, 127-134.

Dobrushin R. L., Fritz J. (1977), Non-equilibrium dynamics of one-dimensional infinite particle systems with a hard-rod interaction. Comm. Math. Phys., **55**, no 3, 275-292.

Dobrushin R. L., Sinai Ya. G., Suhov Yu. M. (1985), Dynamical systems in statistical mechanics. In Modern Mathematical Problems, Dynamical Systems, **2**(in Russian). 235-284, In Encyclopaedia of Math. Sci., Dynamical Systems, **2**,(1989)(Engl. transl.), Springer-Verlag, Berlin.

Dobrushin R., Sokolovskii F. (1991), Higher order hydrodinamical equations for a system of independent random walks. In Random walks, Brownian motion and Interacting Particle Systems, A Festschrift in Honor of Frank Spitzer, Birkhauser, Boston - Basel - Berlin, 231-254.

Dobrushin R. L., Suhov Yu. M. (1979), Time asymptotics for some degenerate models of infinite particles evolution system. In Modern Mathematical Problems (in Russian) **14**, 148-254.

Doob J. (1953), Stochastic Processes. John Wiley & Sons, New-York, Chapman & Hall, London, 654p.

Fritz J., Dobrushin R. L. (1977), Non-equilibrium dynamics of two-dimensional infinite particle systems with a singular interaction. Comm. Math. Phys., **57**, No 1, 67-81.

Gallavotti G., Lanford O. E., Lebowitz J. L. (1972), Thermodynamic limit of the time dependent correlation functions for one-dimensional systems. J. Math. Phys., **13**, No. 11, 2898-2905.

Georgii H. O. (1958), Gibbsian Measures and Phase Transition. Walter de Gruyter, Berlin - New York, 525p.

Gurevich B. M., Suhov Yu. M. (1976-1982), Stationary solutions of the Bogolubov hierarchy equations in classical statistical mechanics, 1-4. Comm. Math. Phys., **49** (1976), No. 1, 69-96, **56**(1977), No. 1, 81-96, **56** (1977), No. 3, 225-236, **84** (1972), No. 4, 333- 376.

Lanford O. E. (1968-1969), The classical mechanics of one-dimensional systems of infinetely many particles, 1. An existence theorem. Comm. Math. Phys., **9**, No. 3, 176-191, 2. Kinetic theory. bf 11, No. 4, 257-292.

Morrey C. B. (1955), On the derivation of the equations of hydrodynamics from statistical mechanics. Comm. Pure Appl. Math., 8, No. 2, 279-326.

Presutti E., Pulverenti M., Tirozzi B. (1976), Time evolution of infinite classical systems with singular, long-range, two-body interaction. Comm. Math. Phys., **47**, No. 1, 81-95.

Sinai Ya. G. (1982), Theory of phase-transition: rigorious results. Pergamon Press, Oxford - New York, 150p.

Sinai Yu. G., Suhov Yu. M. (1974), On the existence theorem for the solutions of Bogolubov class of equations, Theor. Math. Phys., **19**, 344-363.

Lecture on

Diffusion Processes on Nested Fractals

by

S. Kusuoka

Introduction.

The diffusion processes on fractals are new objects in the probability theory. The Brownian motion in the Sierpinski gasket was first constructed by Goldstein [9] and the author [14]. Soon after that, Barlow and Perkins [4] did a wonderful research on it, especially on its transition probability density. Also, Kigami [13] studied its infinitesimal generator by analytic approach.

Recently, Lindstrom [16] introduced a notion, "nested fractals", and constructed diffusion processes on them. By virtue of his work, we obtain diffusion processes on various kinds of fractals.

Also, Barlow and Bass [1], [2] studied the diffusion process on Sierpinski Carpet. Nested fractals are finitely ramified fractals, but Sierpinski Carpet is continuously connected fractals and is much harder to study. Recently, Barlow, Bass and Sherwood [3] introduced an excellent idea to study spectral dimension of such comlex fractals.

Sierpinski gaskets are the simplest fractals and have very strong symmetry. Because of that, one can study fine structure of them. For example, Fukushima and Shima [8] determined the structure of the spectrum of the infinitesimal generators (Rammal and Toulouse [19] did a related work before).

Quite recently, Kumagai constructed non-symmetric diffusion processes on Sierpinski gasket. Also, Kigami discovered a nice embedding of finitely ramified fractals into Euclidean spaces such that the restriction of continuously differentiable functions belong to the domain of the Dirichlet form.

As we saw above, "the diffusion processes on fractals" is one of the most active fields in the probability theory, and the objects are increasing rapidly. This lecture note is based on a course of lectures in Nankai Institute of Mathematics, Tianjin. The author gave a course of lectures from May 15th to May 29th in 1989 there. He tried to give an introduction to this new field. The lectures in the first week was on the Brownian motion on Sierpinski gaskets, and the lectures in the second week was on the diffusion processes on nested fractals and their Dirichlet forms. Since the content of the lectures in the first week was a proper subset of published papers, the author omited it from the present lecture note. However, the author tried to make this lecture note self-contained.

The author thanks to Professor Chen Mu-Fa who invited him to China and gave a chance to give lectures. Also, he thanks to the audience of these lectures in China, in particular, Professors Wu Rong, Yan Shi-Jian, Yan Jia-an and Zhou Xian-Yin for useful suggestions.

1. Self-similar Fractal.

(1.1)**Defintinion.** *Let* $\alpha > 0$. *We say that* $\psi:\mathbb{R}^D\to\mathbb{R}^D$ *is an* α*-similitude, if* $|\psi(x)-\psi(y)| = \alpha^{-1}|x-y|$ *for any* $x,y \in \mathbb{R}^D$.

Then the following is obvious because α-similitudes are contraction maps if $\alpha > 1$.

(1.2)**Proposition.** *If* $\alpha > 1$ *and* $\psi:\mathbb{R}^D\to\mathbb{R}^D$ *is an* α*-similitude,* ψ *has a unique fixed point.*

From now on, we let $\alpha > 1$, and $\{\psi_0,\dots,\psi_N\}$ is a family of α-similitudes on $\mathbb{R}^D$. Let $\Omega = \{0,1,\dots,N\}^{\mathbb{N}}$. We think $\{0,1,\dots,N\}$ is a topological space with discrete topology. Then Ω becomes a metrizable topological space.

(1.3)**Proposition.** *For any* $\omega \in \Omega$ *and* $x \in \mathbb{R}^D$,

$\lim_{n\to\infty} \psi_{\omega(1)}(\psi_{\omega(2)}(\dots(\psi_{\omega(n)}(x))\dots)$ *exists and is indehendent of* x.

Proof. Let a_i be a fixed point of ψ_i, $i=0,1,\dots,N$. Then, since $|\psi_i(y)-a_i| = \alpha^{-1}|y-a_i|$, we have

$$|\psi_i(y)| \leqq \alpha^{-1}|y| + 2|a_i|,\ y \in \mathbb{R}^D.$$

Let $r = 2(1-\alpha^{-1})^{-1}\sum_{j=0}^{n}|a_j|$, and let $B_0 = \{y\in\mathbb{R}^D;\ |y| \leqq r\}$. Then we see that $\psi_i(B_0) \subset B_0$, $i=0,\dots,n$.

Let $B_n = \psi_{\omega(1)}(\dots(\psi_{\omega(n)}(B_0))\dots)$. Then we see that

$$\begin{aligned} B_{n+1} &= \psi_{\omega(1)}(\dots(\psi_{\omega(n)}(\psi_{\omega(n+1)}(B_0)))\dots) \\ &\subset \psi_{\omega(1)}(\dots(\psi_{\omega(n)}(B_0))\dots) = B_n. \end{aligned}$$

Also, we see form the definition of α-similitudes that B_n is a ball of radius $\alpha^{-n}r$ with a center $y_n = \psi_{\omega(1)}(\dots(\psi_{\omega(n)}(0))\dots)$. Since $y_{n+1}\in B_n$, we have $|y_{n+1}-y_n| \leqq \alpha^{-n}r$. Thus we see that $\lim_{n\to\infty} y_n$ exists.

Also, we see that $|\psi_{\omega(1)}(\dots(\psi_{\omega(n)}(x))\dots)-y_n| \leqq \alpha^{-n}|x|$, $n\geqq 1$. Therefore we obtain our assertion.

Let us define a map $\pi:\Omega\to\mathbb{R}^D$ by $\pi(\omega) = \lim_{n\to\infty} \psi_{\omega(1)}(\ldots(\psi_{\omega(n)}(0))\ldots)$.

Then we have the following.

(1.4)**Proposition.** *$\pi:\Omega\to\mathbb{R}^D$ is continuous.*

Proof. Let B_0 be a ball in $\mathbb{R}^D$ as in the proof of Proposition(1.3). Then we see that $\pi(\Omega) \subset B_0$. Assume that ω_1, $\omega_2 \in \Omega$ and $\omega_1(k) = \omega_2(k) = i_k$, $k=1,\ldots,n$. Then we see that

$$\pi(\omega_j) \in \psi_{i_1}(\ldots(\psi_{i_n}(\pi(\Omega))\ldots) \subset \psi_{i_1}(\ldots(\psi_{i_n}(B_0))\ldots),\ j=1,2.$$

Since $\psi_{i_1}(\ldots(\psi_{i_n}(B_0))\ldots)$ is a ball of radius $\alpha^{-n}r$, we see that

$$|\pi(\omega_1)-\pi(\omega_2)| \leqq 2\alpha^{-n}r.$$

This proves the coninuity of π.

We use the following notion throughout this paper.

For any ω, $\omega' \in \Omega$ and $n \geqq 1$, $[\omega,\omega']_n$ denotes an element of Ω given by

$$[\omega,\omega']_n(k) = \begin{cases} \omega(k), & k \leqq n \\ \omega'(k-n), & k \geqq n+1 \end{cases}.$$

For $n \geqq 1$ and $i_1,\ldots,i_n \in \{0,\ldots,N\}$, $\langle i_1\ldots i_n\rangle$ denotes an element of Ω given by

$$\langle i_1\ldots i_n\rangle(k) = \begin{cases} i_k, & k \leqq n \\ i_n, & k \geqq n+1 \end{cases}.$$

T denotes a map in Ω given by $(T\omega)(k) = \omega(k+1)$, $k\geqq 1$. ν denotes a probability measure on Ω such that $\{\omega(n);\ n=1,\ldots\}$ are independently identically distributed under ν and $\nu(\omega(1)=i) = \frac{1}{N+1}$, $i\in\{0,\ldots,N\}$.

Then following is obvious.

(1.5)**Proposition.** (1) *For any* $\omega \in \Omega$, $n \geqq 1$, *and* $i_1,\ldots,i_n\in\{0,\ldots,N\}$

$$\psi_{i_1}(\ldots(\psi_{i_n}(\pi(\omega)))\ldots) = \pi([\langle i_1\ldots i_n\rangle,\omega]_n).$$

(2) $\psi_{\omega(1)}(\pi(T\omega)) = \pi(\omega)$, $\omega \in \Omega$.

Now we have the following.

(1) $\bigcup_{i=0}^{N} \psi_i(V) \subset V$, *and*

(2) $\psi_i(V)\cap\psi_j(V) = \phi$, *for any* $i,j\in\{0,\ldots,N\}$ *with* $i\neq j$.

(1.8)**Proposition.** *If* $\{\psi_0,\ldots,\psi_N\}$ *satisfies the open set condition, then there is an* $M \in \mathbb{N}$ *such that* $\#(\pi^{-1}(x)) \leq M$ *for all* $x \in \mathbb{R}^D$.

Proof. Let V be as in Definition(1.7). Then we have $\bigcup_{i=0}^{N} \psi_i(\overline{V}) \subset \overline{V}$. Here $\overline{V}$ is the closure of V. So by Proposition(1.6)(2), we see that $E = \pi(\Omega) \subset \overline{V}$. Let $B_n = \{x\in\mathbb{R}^D;\ |x|\leq\alpha^{-n}\}$, $n\geq 0$. Then we see that $(B_0+x)\cap V\neq\phi$ for any $x\in E$. Therefore if we let $g(x) = |(B_0+x)\cap V|$, $x\in\mathbb{R}^D$, then $g:\mathbb{R}^D\to[0,\infty)$ is continuous and $g(x) > 0$, $x\in E$. Here $|B|$ denotes the volume of the set B. So we see that

$\delta = \inf\{g(x);\ x\in E\} > 0$.

Suppose that $\omega_1,\ldots,\omega_m \in \pi^{-1}(x)$, $m \geq 1$ and $\omega_1,\ldots,\omega_m$ are distinct. Then $x\in E$. Moreover, $x \in \psi_{\omega_i(1)}(\ldots(\psi_{\omega_i(n)}(E))\ldots)$, $i=1,\ldots,m$. So there are $y_{i,n}\in E$, $i=1,\ldots,m$, $n\geq 1$, such that $x = \psi_{\omega_i(1)}(\ldots\psi_{\omega_i(n)}(y_{i,n}))\ldots)$. Then we see that

$$x+B_n = \psi_{\omega_i(1)}(\ldots(\psi_{\omega_i(n)}(y_{i,n}+B_0)\ldots),$$

and so

$$\bigcup_{i=1}^{m} \psi_{\omega_i(1)}(\ldots(\psi_{\omega_i(n)}((y_{i,n}+B_0)\cap V))\ldots) \subset x+B_n .$$

By the open set condition, if n is sufficiently large, $\psi_{\omega_i(1)}(\ldots\psi_{\omega_i(n)}(V)\ldots)$, $i=1,\ldots,m$, are mutually disjoint. Therefore if n is sufficiently large,

$$\sum_{i=1}^{m} |\psi_{\omega_i(1)}(\ldots(\psi_{\omega_i(n)}((y_{i,n}+B_0)\cap V))\ldots)| \leq |x+B_n|,$$

and so

$$\sum_{i=1}^{m} |(y_{i,n}+B_0)\cap V| \leq |B_0|.$$

This implies that $m\delta \leq |B_0|$. Therefore letting $M = [|B_0|/\delta]$, we

(1.6)**Theorem.** (1) *$E = \pi(\Omega)$ is a compact set satisfying*

$$\bigcup_{i=0}^{N} \psi_i(E) = E.$$

(2) *If C is a non-empty closed set satisfying* $\bigcup_{i=0}^{N} \psi_i(C) \subset C$, *then* $\pi(\Omega) \subset C$.

(3) *If K is a non-empty compact set satisfying* $\bigcup_{i=0}^{N} \psi_i(K) \supset K$, *then* $K \subset \pi(\Omega)$. *In particular, if K is a non-empty compact set satisfying* $\bigcup_{i=0}^{N} \psi_i(K) = K$, *then* $K = \pi(\Omega)$.

Proof. (1) Since Ω is compact and π is continuous, $\pi(\Omega)$ is compact. Also, we see that

$$\begin{aligned}\bigcup_{i=0}^{N} \psi_i(\pi(\Omega)) &= \bigcup_{i=0}^{N} \{\psi_i(\pi(\omega));\ \omega\in\Omega\} \\ &= \pi(\bigcup_{i=0}^{N} \{[\langle i\rangle,\omega]_1;\ \omega\in\Omega\} \\ &= \pi(\Omega).\end{aligned}$$

This proves our assertion (1).

(2) Let $x_0 \in C$. Then $\psi_{\omega(1)}(\ldots(\psi_{\omega(n)}(x_0))\ldots) \in C$, $\omega\in\Omega$, $n\geqq1$. Since C is clsoed and $\pi(\omega) = \lim_{n\to\infty} \psi_{\omega(1)}(\ldots(\psi_{\omega(n)}(x_0))\ldots)$, we see that $\pi(\Omega) \subset C$.

(3) Since K is compact, there is an $r > 0$ such that $K \subset B = \{y\in\mathbb{R}^D;\ |y|\leqq r\}$. Let $x \in K$. Then by the assumption, there is an $i_1\in\{0,\ldots,N\}$ such that $x \in \psi_{i_1}(K)$. Similary, we see that there are $i_k\in\{0,\ldots,N\}$, $k\geqq1$, such that $x \in \psi_{i_1}(\ldots(\psi_{i_n}(K))\ldots)$ for any $n \geqq 1$. Then $|x-\psi_{i_1}(\ldots(\psi_{i_n}(0))\ldots)| \leqq \alpha^{-n}r$. Let ω be an element of Ω such that $\omega(k) = i_k$, $k\geqq1$. Then we see that $x = \pi(\omega)$. Therefore $K \subset \pi(\Omega)$. The final assertion is obvious from the assertion (2).

(1.7)**Definition.** *We say that $(\psi_0,\ldots,\psi_N)$ satisfies the open set condition, if there exists a non-void open set V such that*

have our assertion.

(1.9)**Corollary.** *If $\{\psi_0, \ldots, \psi_N\}$ satisfies the open set condition, the fixed points of ψ_i's are distinct.*

Proof. Suppose that $i \neq j$ and the fixed points of ψ_i and ψ_j are the same. Let x be the fixed point of ψ_i. Then we see that $[\langle i \rangle, \langle j \rangle]_n \in \pi^{-1}(x)$, n=1,2,.... This contradicts to Proposition(1.8).

The following is a corollary to the results by Hutchinson [12]. Since we will not use this result, we will not give its proof.

(1.10)**Theorem.** *Suppose that $\{\psi_0, \ldots, \psi_N\}$ is a family of α-similitudes in $\mathbb{R}^D$ satisfying the open set condition. Then the Hausdorff dimension d_f of $E = \pi(\Omega)$ is $\log(N+1)/\log \alpha$. Moreover, d_f-Hausdorff measure on E and $\nu \circ \pi^{-1}$ are the same up to a constant factor.*

Now we give several examles.

Example 1 (Sierpinski Gasket). Let $D \geqq 2$ and $\{x_1, \ldots, x_D\} \subset \mathbb{R}^D$ such that $|x_i| = 1$, i=1,...,D, and $|x_i - x_j| = 1$, $i \neq j$. Let $\psi_i : \mathbb{R}^D \to \mathbb{R}^D$, i=0,1,...,D, be given by

$$\psi_0(x) = \frac{1}{2} \cdot x$$

and

$$\psi_i(x) = \frac{1}{2}(x - x_i) + x_i, \quad i=1,\ldots,D.$$

Then $\{\psi_0, \ldots, \psi_D\}$ is a family of 2-similitudes satisfying the open set condition. In this case, N = D and the Hausdorff dimension d_f of E $= \pi(\Omega)$ is $\log(D+1)/\log 2$.

Example 2 (Sierpinski Carpet). Let $D \geqq 2$ and $\tilde{A} = \{0, \frac{1}{2}, 1\}^D \subset \mathbb{R}^D$. Then $\#(\tilde{A}) = 3^D$. Let $A = \tilde{A} \setminus \{(\frac{1}{2}, \ldots, \frac{1}{2})\}$ and $\psi_a : \mathbb{R}^D \to \mathbb{R}^D$, $a \in A$, be given by $\psi_a(x) = \frac{1}{3}(x-a) + a$, $x \in \mathbb{R}^D$. Then $\{\psi_a;\ a \in A\}$ is a family of 3-similitudes in $\mathbb{R}^D$ satisfying the open set condition. In this case, $N = 3^D-2$, and the Hausdorff dimension d_f of $E = \pi(\Omega)$ is $\log(3^D-1)/\log 3$.

Example 3 (snowflake fractal). Let $D = 2$ and $N = 6$. Let $a_k=(\cos(k\pi/3),\sin(k\pi/3))$, $k=1,\ldots,6$. Let $\psi_k:\mathbb{R}^D \to \mathbb{R}^D$, $k=0,\ldots,6$, be given by

$$\psi_0(x) = \frac{1}{3}\cdot x$$

and

$$\psi_k(x) = \frac{1}{3}\cdot(x-a_k)+a_k\ ,\quad k=1,\ldots,6.$$

Then $\{\psi_k, k=0,\ldots,6\}$ is a family of 3-similitudes on $\mathbb{R}^2$ satisfying open set condition and the Hausdorff dimension d_f of $E = \pi(\Omega)$ is $\log 7/\log 3$.

(1.11)**Remark.** Suppose that $\psi_0(x) = \alpha^{-1}x$, $x \in \mathbb{R}^D$. Then it is easy to see that $\pi([\langle 0\rangle,\omega]_n) = \alpha^{-n}\pi(\omega)$, $\omega \in \Omega$, $n \geqq 1$. Let $\Omega_0=\{\omega \in \{0,1,\ldots,N\}^{\mathbb{Z}};$ there is an $n \in \mathbb{Z}$ such that $\omega(k)=0$ for all $k \leqq n\}$. Also, we define a metric function $d_{\Omega_0}:\Omega_0 \times \Omega_0 \to [0,\infty)$ by

$$d_{\Omega_0}(\omega_1,\omega_2) = \sum_{k=-\infty}^{\infty} 2^{-k}\gamma(\omega_1(k),\omega_2(k)),\quad \omega_1,\omega_2 \in \Omega.$$

Here $\gamma(i,j) = \begin{cases} 0 & \text{if } i=j \\ 1 & \text{if } i \neq j \end{cases}$.

Then (Ω_0, d_{Ω_0}) is a Polish space. Let $S:\Omega_0 \to \Omega_0$ be given by $(S\omega)(k) = \omega(k+1)$ and $P:\Omega_0 \to \Omega$ be given by $(P\omega)(k) = \omega(k)$, $k \in \mathbb{N}$. Then it is easy to see that $\alpha^n\pi(P(S^{-n}\omega))$ is independent of n, if n is sufficiently large. So if we define $\tilde{\pi}:\Omega_0 \to \mathbb{R}^D$ by $\tilde{\pi}(\omega) = \lim_{n\to\infty} \alpha^n\pi(P(S^{-n}\omega))$, $\omega \in \Omega$, then $\tilde{\pi}$ is continuous. Moreover, $\alpha\cdot\tilde{\pi}(\Omega_0) = \pi(\Omega_0)$. If $\{\psi_0,\ldots,\psi_N\}$ satisfies the open set condition, the Hausdorff dimension of $\tilde{\pi}(\Omega_0)$ is $\log(N+1)/\log \alpha$.

2. Nested fractals and their geometrical properties.

Let $\alpha > 1$ and $\{\psi_0,\ldots,\psi_N\}$ be an α-similitudes in $\mathbb{R}^D$. We will impose several assumptions on this family $\{\psi_0,\ldots,\psi_N\}$.

First we assume that

(A-0) $\{\psi_0,\ldots,\psi_N\}$ satisfies the open set condition.

Let F_0 be the set of fixed points of ψ_i's, $i=0,\ldots,N$. Then by Corollary(1.9), we see that $\#(F_0) = N+1$.

(2.1)**Definition.** *An essential fixed point is an element* x *of* F_0 *such that there are* $i,j \in \{0,\ldots,N\}$, $i\neq j$, *and* $y\in F_0$ *for which* $\psi_i(x)=\psi_j(y)$. *We denote by* F *the set of essential fixed points.*

Now we introduce several notions. For any subset A in $\mathbb{R}^D$ and $i_1,\ldots,i_n\in\{0,\ldots,N\}$, $A_{i_1\ldots i_n}$ denotes the set $\psi_{i_1}(\ldots\psi_{i_n}(A)\ldots)$.
Also, we let $F^{(n)} = \bigcup_{i_1,\ldots,i_n=0}^{N} F_{i_1\ldots i_n}$ for each $n \geqq 1$, and $F^{(0)} = F$.
An element of $F^{(n)}$ is called an n-point for each $n \geqq 0$. For each $n \geqq 1$, a set of the form $F_{i_1\ldots i_n}$ is called an n-cell, and a set of the form $E_{i_1\ldots i_n}$ is called an n-complex. 0-cell is the set F, and 0-complex is the set E.

Now we assume the following.

(A-1)(**connectivity**) For any two 1-cells C and C', there is a sequence $\{C_i;\ i=0,\ldots,n\}$ of 1-cells such that $C_0 = C$, $C_n = C'$ and $C_{i-1}\cap C_i\neq\phi$ $i=1,\ldots,n$.

For any $x,y\in\mathbb{R}^D$ with $x\neq y$, H_{xy} denotes the hyperplane given by

$$H_{xy} = \{z\in\mathbb{R}^D;\ |z-x| = |z-y|\},$$

and U_{xy} denotes the reflection with respect to H_{xy}, i.e.,

$$U_{xy}z = z - 2\cdot|x-y|^{-2}(z,x-y)_{\mathbb{R}^D}\cdot(x-y),\quad z\in\mathbb{R}^D.$$

We assume the following two assumptions furthermore.

(A-2)(symmetry) If $x,y \in F$ and $x \neq y$, then U_{xy} maps n-cells to n-cells, and maps any n-cell which contains elements in both sides of H_{xy} to itself for each $n \geqq 0$.

(A-3)(nesting) If $n \geqq 1$, and if $(i_1,\ldots,i_n)$ and $(j_1,\ldots,j_n)$ are distinct elements of $\{0,1,\ldots,N\}^n$, then

$$E_{i_1\ldots i_n} \cap E_{j_1\ldots j_n} = F_{i_1\ldots i_n} \cap F_{j_1\ldots j_n} .$$

(2.2)**Definition.** *A nested fractal is a self-similar fractal* E *associated with* α*-similitudes* $\{\psi_0,\ldots,\psi_N\}$ *satisfying the assumptions* (A-0) ~ (A-3) *and that* $\#(F) \geqq 2$.

(2.3)**Remark.** *Sierpinski Gasket (Example* 1) *and snowflake fractal (Example* 3) *are nested fractals, but Sierpinski Carpet (Example* 2) *is not.*

We say that a map U in $\mathbb{R}^D$ is a symmetry, if $|U(x)-U(y)| = |x-y|$ for any $x,y \in \mathbb{R}^D$ and if U maps n-cells to n-cells for each $n \geqq 0$. It is easy to see that the set of symmetries becomes a group, i.e., if U_1 and U_2 are symmetry, U_1U_2 and U_1^{-1} are symmetries.

(2.4)**Lemma.** *If* $x,y,x',y' \in F$ *and* $|x-y| = |x'-y'|$, *then there is a symmetry* U *such that* $U(x) = x'$ *and* $U(y) = y'$.

Proof. If $y=y'$, $U = U_{xx'}$ is a desired one. Assume that $y \neq y'$. Let $x'' = U_{yy'}(x')$. If $x = x''$, we can take $U_{yy'}$ as U. So assume that $x'' \neq x$. Since $|x-y| = |x'-y'| = |x''-y|$, we have $U_{xx''}(y) = y$. Let $U = U_{xx''}U_{yy'}$. Then we have $U(y') = U_{xx''}(y) = y$ and $U(x') = U_{xx''}(x'') = x$. This proves our assertion.

Let $\delta_0 = \min\{|x-y|;\ x,y \in F,\ x \neq y\}$, and $\delta_n = \alpha^{-n}\delta_0$, $n \geqq 1$. Then it is easy to see that $\delta_n = \min\{|x-y|;\ x,y \in C,\ x \neq y\}$ for any n-cell C.

(2.5)**Definition.** *We say that* $x, y \in E$ *are* n*-neighbors, if there is an* n*-cell for which* $x, y \in C$. *We say that* $x, y \in E$ *are nearest* n*-neighbors, if* x, y *are* n*-neighbors and* $|x-y| = \delta_n$. *We say that* $s_1, \ldots, s_m$ *is an* n*-walk, if* s_i *and* s_{i+1} *are* n*-neighbors for all* $i = 1, \ldots, m-1$. *We say that* $s_1, \ldots, s_m$ *is a strict* n*-walk, if* s_i *and* s_{i+1} *are nearest* n*-neighbors for all* $i = 1, \ldots, m-1$.

(2.6)**Lemma.** *If* $x, y \in F$ *and* $x \neq y$, *then there is a strict* 0*-walk* $s_1, \ldots, s_m$ *such that* $s_1 = x$ *and* $s_m = y$.

Proof. First, note that each point $x \in F$ has a nearest 0-neighbor. This is because there is a pair $y, z \in F$ with $|y-z| = \delta_0$, and so $U_{yx}(z) \in F$ and $|x-U_{yx}(z)| = \delta_0$.

Now let $A = \{(x,y) \in F \times F;\ x \neq y,\ x$ and y cannot be connected by strict 0-walk$\}$, and assume that $A \neq \phi$. Let $\varepsilon = \min\{|x-y|;\ (x,y) \in A\} > 0$. Then $\varepsilon > \delta_0$. Assume that $(x,y) \in A$ and $|x-y| = \varepsilon$. Let z be an nearest 0-neighbor of x. Then $(z,y) \in A$, and so $|z-y| \geqq |x-y|$. Let $x' = U_{zy}(x)$. Then $|x'-y| = |U_{zy}(x)-U_{zy}(z)| = |x-z| = \delta_0$. So $(x,x') \in A$ and $|x-x'| \geqq |x-y|$.

Let $r = |y-z|$ and $e = r^{-1}(y-z)$. Then we have

$|(e,x-x')| = |x-x'|$, and

$|(e,y-z)| = |y-z|$.

Also, we have

$|(e,x-z)| \leqq |x-z| = \delta_0 < |y-z| = |(e,y-z)|$, and

$|(e,y-x)| \leqq |x-y| \leqq |y-z| \leqq |(e,y-z)|$.

If $|(e,y-x)| = |y-z| = r$, then $y-x = \pm r \cdot e$. But this implies that $x=z$ or $y = \frac{1}{2}(x+z)$, and so this contradicts $|x-z| = \delta_0$. Therefore we see that

$|(e,y-x)| < |(e,y-z)|$.

Since $(e,y-z) = (e,y-x) + (e,x-z)$, we see that $(e,y-z)$, $(e,y-x)$ and $(e,x-z)$ have the same sign and they are non-zero. Observe that

$(e,x'-y) = (e, U_{zy}(x)-U_{zy}(z)) = -(e,x-z)$, and so

$(e,x'-x) = (e,x'-y) + (e,y-x) = -(e,x-z) + (e,y-x)$.

Therefore $|(e,x'-x)| < |(e,y-x)|$, but this implies that $|x'-x| < |y-x|$. This is the contradiction. So we see that $A = \phi$.

This completes the proof.

(2.7)**Lemma.** *Let* x,y,z *are distinct points in* F. *Then there is a strict* 0-*walk which connects* x *and* y *and avoids* z.

Proof. For each $(x,y) \in F\times F$, let

$d(x,y) = \min\{ m \geqq 1;\ s_1,\dots,s_m$ is a strict 0-walk connecting x and y$\}$ if $x \neq y$,

and

$d(x,y) = 0$, if $x = y$.

Then d is a metric function on F.

Now let $A = \{(x,y,z)\in F^3;$ x,y,z are distinct and there is no strict 0-walk which connects x and y and avoids z$\}$, and assume that $A \neq \phi$. Let $d = \max\{d(x,y);\ (x,y,z)\in A\}$. Take $(x,y,z)\in A$ such that $d(x,y) = d$, and fix them. If $u\in F\setminus\{y\}$ and $(x,u,y)\in A$, then $d(x,u) > d(x,y)$ and so this contradicts our assumption. So we see that $(x,u,y)\in F^3\setminus A$ for any $u\in F\setminus\{y\}$.

Let $s_1,\dots,s_d$ is a strict 0-walk connecting x and y. Then from our assumption $z=s_k$ for some $k\in\{2,\dots,d-1\}$. Let $s_i' = U_{yz}(s_i)$, $i=1,\dots,d$. Then $s_1',\dots,s_k'$ is a shortest strict 0-walk connecting $x'=U_{yz}(x)$ and $y=U_{yz}(z)$. Note that $x'\neq y$ and $s_i'\neq z$, $i=1,\dots,k$. Then $(x,x',y)\in F^3\setminus A$, and so there is a strict 0-walk $t_1,\dots,t_r$ which connects x and x' and avoids y.

If $\{t_1,\dots,t_r\}$ does not contain z, then $t_1,\dots,t_{r-1},s_1',\dots,s_k'$ is a strict 0-walk which connects x and y and avoids z. This contradicts to our assumption. On the other hand, suppose that $t_j=z$ for some $j \in\{1,\dots,k\}$. Then $t_r,\dots,t_j$ is a strict 0-walk which connects x' and z and avoids y. Then $U_{yz}(t_r),\dots,U_{yz}(t_j)$ is a strict 0-walk which connects x and y and avoids z. So we have a contradiction again. This proves our lemma.

(2.8)**Lemma.** *Each element in* F *belongs to only one* n-*cell for each* $n \geq 0$.

Proof. Let $x \in F$. Then x is a fixed point of ψ_i for some $i \in \{0,1,\dots,N\}$. So $x \in F_{i\dots i}$. Suppose that x belongs to another n-cell $F_{j_1\dots j_n}$. Then there is a $k \in \{0,\dots,N\}$ such that $x = \psi_{j_1}(\dots(\psi_{j_n}(y)\dots)$ for the fixed point y of ψ_k. Then we see that $[\langle i\rangle,\langle j_1\dots j_n k\rangle]_m \in \pi^{-1}(x)$, for any $m \geq 0$, and so $\#(\pi^{-1}(x)) = \infty$. This contradicts Proposition(1.8).

(2.9)**Proposition.** *Any* 1-*cell contains at most one element of* F.

Proof. Suppose that F_i contains two elements $x,y \in F$. Since fixed points of $\psi_0,\dots,\psi_N$ are distinct, one of them, say x, is not a fixed point of ψ_i. So x is a fixed point of ψ_j for some $j \in \{0,\dots,N\}$. Then $x \in F_i \cap F_j$. This contradicts to Lemma(2.8).

This completes the proof.

(2.10)**Proposition.** *Let* $x,y \in F^{(1)}$. *Then there is a strict* 1-*walk* $s_1,\dots,s_n$ *such that* $s_1 = x$, $s_n = y$ *and* $s_k \in F^{(1)} \setminus F$, $k=2,\dots,n-1$.

Proof. By the assumption (A-1) of connectivity, there are distinct 1-cells $C_1,\dots,C_n$ such that $x \in C_1$, $y \in C_n$ and $C_i \cap C_{i+1} \neq \emptyset$, $i=1,\dots,n-1$. Let $t_i \in C_i \cap C_{i+1}$, $i=1,\dots,n-1$. Then by Proposition(2.9), $t_i \in F^{(1)} \setminus F$. Let $t_0 = x$ and $t_n = y$. Since C_i is just a copy of F and C_i contains at most one element of F by Proposition(2.8), there is a strict 1-walk in C_i which connects t_{i-1} and t_i and avoids an element of F for each $i = 1,\dots,n$. So connecting them, we obtain our desired one.

Let $I = \{i \in \{0,\dots,N\}$; the fixed point of ψ_i is essential fixed point$\}$. Then $\#(I) = \#(F)$ and $\{\pi(\langle i \rangle); i \in I\} = F$. For each $n \geq 0$, let $A_n = \{\omega \in \Omega;\ \omega(k) = \omega(n+1),\ k \geq n+1,\ \omega(n+1) \in I\}$. Then it is easy to see that $\pi(A_n) = F^{(n)}$.

(2.11)**Definition.** *We define an equivalence relation* $\sim$ *on* $\{0,1,\dots,N\} \times I$ *by the following.*

$$(k,i) \sim (k',i'),\quad (k,i),(k',i') \in \{0,\dots,N\} \times I,$$

$$\textit{if}\ \psi_k(\pi(\langle i \rangle)) = \psi_{k'}(\pi(\langle i' \rangle)).$$

(2.12)**Proposition.** *Let* $(k,i),(k',i') \in \{0,\dots,N\} \times I$. *If* $(k,i) \neq (k',i')$ *and* $(k,i) \sim (k',i')$, *then* $k \neq k'$, $k \neq i$ and $k' \neq i'$.

Proof. First, assume that $k=i$. Then $\psi_k(\pi(\langle i \rangle)) = \pi(\langle i \rangle) \in F$ and $\pi(\langle i \rangle) \in F_{ki} \cap F_{k'i'}$. This contradicts Proposition(2.7). So we have $k \neq i$ and also $k' \neq i'$. Assume that $k=k'$. Then we have $\psi_k(\pi(\langle i \rangle)) = \psi_k(\pi(\langle i' \rangle))$, and so $\pi(\langle i \rangle) = \pi(\langle i' \rangle)$. This implies $i=i'$. This contradicts our assumption.

This completes the proof.

(2.13)**Proposition.** *Suppose that* $\omega_1, \omega_2 \in \Omega$, $\omega_1 \neq \omega_2$, *and* $\pi(\omega_1) = \pi(\omega_2)$. *Then there is an* $n \geq 1$ *such that*

$$\omega_1(k) = \omega_2(k),\ k \leq n-1,$$

$$\omega_1(k) = \omega_1(n+1) \in I,\ k \geq n+1,$$

$$\omega_2(k) = \omega_2(n+1) \in I,\ k \geq n+1,$$

and

$$(\omega_1(n), \omega_1(n+1)) \sim (\omega_2(n), \omega_2(n+1)).$$

Proof. Let $n = \min\{k \geq 1;\ \omega_1(k) \neq \omega_2(k)\}$, and $\ell_k = \omega_1(k) = \omega_2(k)$, $k \leq n-1$. Then we see that $\pi(T^{n-1}\omega_j) = \psi_{\ell_{n-1}}^{-1}(\dots(\psi_{\ell_1}^{-1}(\pi(\omega_j)))\dots)$, $j=1,2$, and so $\pi(T^{n-1}\omega_1) = \pi(T^{n-1}\omega_2)$. Note that $\pi(T^{n-1}\omega_j) \in E_{\omega_j(n)}$, $j=1,2$, and

$\omega_1(n) \neq \omega_2(n)$. Therefore $\pi(T^{n-1}\omega_j) \in E_{\omega_1(n)} \cap E_{\omega_2(n)} = F_{\omega_1(n)} \cap F_{\omega_2(n)}$. So we have $\pi(T^n\omega_j) = \psi_{\omega_j(n)}{}^{-1}(\pi(T^{n-1}\omega_j)) \in F$, $j=1,2$. Then by Proposition(2.8), we see that $T^n\omega_j = \langle i_j \rangle$ for some $i_j \in I$, $j=1,2$. Then we have $\psi_{\omega_1(n)}(\pi(\langle\omega_1(n+1)\rangle)) = \psi_{\omega_2(n)}(\langle\omega_2(n+1)\rangle)$.

These imply our assertion.

The following is an easy consequence of Proposition(2.13).

(2.14)**Corollary.** (1) *If* $x \in E \setminus (\bigcup_{n=0}^{\infty} F^{(n)})$, *then* $\#(\pi^{-1}(x)) = 1$.

(2) *If* $x \in F^{(n)} \setminus F^{(n-1)}$, $n \geqq 0$, *and* $\pi^{-1}(x) = \{\omega_1, \ldots, \omega_m\}$, *then*

$$\omega_1(k) = \omega_2(k) = \ldots = \omega_m(k), \ k \leqq n-1,$$

$$\omega_j(k) = \omega_j(n+1) \in I, \ k \geqq n+1, \ j=1,\ldots,m,$$

and

$$(\omega_j(n), \omega_j(n+1)) \sim (\omega_\ell(n), \omega_\ell(n+1)), \ j,\ell=1,\ldots,m.$$

3. Transition probability of Markov chain.

Let $r = \#\{|x-y|;\ x,y\in F,\ x\neq y\}$, and let $\ell_1,\ldots,\ell_r$ be such that $0 < \ell_1 <\ldots< \ell_r$ and $\{\ell_1,\ldots,\ell_r\} = \{|x-y|; x,y\in F,\ x\neq y\}$. Now let $C_i = \{(x,y)\in F^2; |x-y|=\ell_i\}$, $i=1,\ldots,r$. Let $m_i(x) = \#\{y\in F;\ (x,y)\in C_i\}$, $i=1,\ldots,r$. Then, because of the symmetry, we see that $m_i(x)$ is independent of $x\in F$. So we write m_i for $m_i(x)$. Let $\mathcal{P} = \{(p_1,\ldots,p_r);\ p_1\geqq p_2\geqq\ldots\geqq p_r,\ \sum_{i=1}^{r} m_i p_i = 1\}$. Obviously, $\mathcal{P}$ is a compact convex set in $\mathbb{R}^r$. For each $x \in F^{(1)}$, we let

$$\rho(x) = \#\{C;\ C \text{ is a 1-cell containing } x\}.$$

Also, for $x,y \in F$, we let

$$\tilde{\rho}(x,y) = \#\{C;\ C \text{ is a 1-cell containing both of } x \text{ and } y\}.$$

Now we define $P:F^{(1)}\times F^{(1)}\times\mathcal{P}\to[0,1]$ by

$$P(x,y;p) = \begin{cases} \rho(x)^{-1}\tilde{\rho}(x,y)p_i & \text{if } \tilde{\rho}(x,y) \geqq 1 \text{ and } |x-y| = \alpha^{-1}\ell_i \\ 0 & \text{if } x=y \text{ or } \tilde{\rho}(x,y)=0 \end{cases}.$$

Here $p = (p_1,\ldots,p_r)$. Then we see that

$$\sum_{y\in F^{(1)}} P(x,y;p) = 1\ ,\quad x\in F^{(1)},\ p\in\mathcal{P}.$$

For each $p\in\mathcal{P}$, let us define probability measures $\{P_x;x\in F^{(1)}\}$ on $W_1 = \{w:\{0,1,\ldots\}\to F^{(1)}\}$ by

$$P_x[\,w(0)=x_0,\ldots,\ w(n) = x_n] = \begin{cases} \prod_{i=1}^{n} P(x_{i-1},x_i;p) & \text{if } x_0 = x \\ 0 & \text{if } x_0 \neq x \end{cases}.$$

Then $\{P_x;\ x\in F^{(1)}\}$ is a function of p. So we sometimes write $P_x^{(p)}$ for P_x. It is obvious that $\{P_x;\ x\in F^{(1)}\}$ is a Markov chain.

Let $\tau^x(w) = \min\{\ n\geqq 0;\ w(n) \in F\setminus\{x\}\ \}$, $x\in F^{(1)}$, $w \in W_1$. In this lecture, we take the convention that $\min\phi = \infty$. Note that

(3.1) $p_1 \geqq (\sum_{i=1}^{r} m_i)^{-1} = (\#(F)-1)^{-1}$, if $(p_1,\ldots,p_r) \in \mathcal{P}$.

So we have the following.

(3.2)**Proposition.** *There are* $C \in (0,\infty)$ *and* $\theta \in (0,1)$ *such that*

$$P_x^{(p)}[\ \tau^x(w) \geqq n\ ] \leqq C\cdot\theta^n \qquad \textit{for any } p\in\mathcal{P},\ x \in F^{(1)} \textit{ and } n \geqq 0.$$

In particular, $P_x^{(p)}[\ \tau^x = \infty\ ] = 0$ *for any* $p\in\mathcal{P}$ *and* $x \in F^{(1)}$.

Proof. By Lemma(2.6), for any $x \in F^{(1)}$ and $y \in F$, there is a strict 1-walk $s_1,\dots, s_{m_{xy}}$ which connects x and y. Let $m = \max\{m_{xy};\ x\in F^{(1)}, y\in F\}$. Then we have

$$P_x[\ \tau^z \geqq m\ ] \leqq 1 - (\#(F)-1)^{-m-1} = \theta_1 < 1$$

for any $x,z \in F^{(1)}$ and $p \in \mathcal{P}$.

Then we see that

$$\begin{aligned} &P_x[\ \tau^x \geqq m\cdot(\ell+1)\ ] \\ &= E^{P_x}[\ P_{w(m\cdot\ell)}[\ \tau^x \geqq m\ ],\ \tau^x \geqq m\cdot\ell\ ] \\ &\leqq \theta_1\cdot P_x[\ \tau^x \geqq m\cdot\ell\ ], \quad \ell \geqq 1. \end{aligned}$$

So we have

$$P_x[\ \tau^x \geqq m\cdot\ell\ ] \leqq \theta_1^{\ell}, \quad \ell \geqq 0.$$

This implies our assertion.

By Proposition(3.2), we can define $p_{xy}(p)$, $p\in\mathcal{P}$, $x,y \in F$, by

$$p_{xy}(p) = P_x^{(p)}[\ w(\tau^x) = y\ ].$$

Then we have the following.

(3.3)Proposition. *If* $x,y,x',y' \in F$ *and* $|x-y| = |x'-y'|$, *then* $p_{xy}(p) = p_{x'y'}(p)$ *for any* $p \in \mathcal{P}$.

Proof. By Lemma(2.4), there is a symmetry U such that $Ux = x'$ and $Uy = y'$. Let $\tilde{U}:W_1\to W_1$ be given by $(\tilde{U}w)(n) = Uw(n)$, $n \geqq 0$. Then it is easy to see that $P_z\circ\tilde{U}^{-1} = P_{Uz}$ for any $z \in F^{(1)}$. Then we have

$$\begin{aligned} p_{x'y'}(p) &= P_{x'}[\ w(\tau^{x'}) = y'\ ] \\ &= P_{x'}[\ (\tilde{U}w)(\tau^x(\tilde{U}w)) = y\ ] \\ &= P_x[\ w(\tau^x) = y\ ] = p_{xy}(p). \end{aligned}$$

By Proposition(3.3), $p_{xy}(p)$, $(x,y)\in C_i$, are the same. So we denote $p_{xy}(p)$, $x,y\in C_i$, by $p_i(p)$ for each $i = 1,\dots,r$. Also, we define $\tilde{p}:\mathcal{P}\to[0,1]^r$ by $\tilde{p}(p) = (p_1(p),\ldots,p_r(p))$.

(3.4)**Proposition.** $\tilde{p}:\mathcal{P}\to[0,1]^r$ *is continuous.*

Proof. Note that

$$P_x^{(p)}[\, w(\tau^x) = y,\ \tau^x \leqq n \,]$$

$$= \sum_{k=1}^{n} \sum_{s_1,\ldots,s_{k-1}\in F^{(1)}\setminus(F\setminus\{x\})} p(x,s_1;p)p(s_1,s_2;p)\ldots p(s_{k-2},s_{k-1};p) \times p(s_{k-1},y;p)$$

So the function $p \to P_x^{(p)}[w(\tau^x)=y,\ \tau^x\leqq n]$ is continuous from $\mathcal{P}$ into $[0,1]$ for any $n \geqq 1$ and $x,y \in F$.

On the other hand, by Proposition(3.2) we have

$$|p_{xy}(p) - P_x^{(p)}[w(\tau^x)=y,\ \tau^x\leqq n]| \leqq C\cdot\theta^n .$$

So we see that $P_x^{(p)}[w(\tau^x)=y,\ \tau^x\leqq n]$ converges to $p_{xy}(p)$ uniformly in $p\in\mathcal{P}$ as $n \to \infty$. This implies our assertion.

The following is the most important observation in Lindstrom [16].

(3.5)**Lemma.** $\tilde{p}(p) \in \mathcal{P}$ *for any* $p \in \mathcal{P}$.

Proof. We prove this lemma in several steps.

Step 1. It is obvious that $\sum_{i=1}^{r} m_i p_i(p) = 1$. So it is sufficient to show that $p_1(p) \geqq p_2(p) \geqq\ldots\geqq p_r(p)$. To prove this, we fix $p\in\mathcal{P}$. Also, we introduce a new Markov chain as follows.

Let $q:F^{(1)}\times F^{(1)}\to[0,1]$ be given by

$$q(x,y) = \begin{cases} \frac{1}{2}\, p(x,y;p) & \text{if } x\neq y \\ \frac{1}{2} & \text{if } x=y \end{cases}$$

Then we see that $\sum_{y\in F^{(1)}} q(x,y) = 1$, $x\in F^{(1)}$. Let $\{Q_x;\ x\in F^{(1)}\}$ be a family of probability measures on W_1 given by

$$Q_x[\, w(0) = x_0,\ldots,\ w(n) = x_n]$$

$$= \begin{cases} \prod_{i=1}^{n} q(x_{i-1},x_i) & \text{if } x_0 = x \\ 0 & \text{if } x_0\neq x \end{cases} .$$

Then $\{Q_x; x\in F^{(1)}\}$ is a Markov chain.

Let $\eta(w) = \min\{n \geqq 1;\ w(n) \neq w(0)\}$, $w\in W_1$. Then It is easy to see that

$Q_x[\eta=\infty] = 0$, $x\in F^{(1)}$, and

$Q_x[w(\eta)=y] = P_x[w(1)=y]$, $x,y\in F^{(1)}$.

Therefore by using the strong Markov property, we see that

$p_{xy}(p) = P_x[w(\tau^x)=y] = Q_x[w(\tau^x)=y]$, $x,y \in F^{(1)}$.

Also, let $\sigma(w) = \min\{n\geqq 1;\ w(n)\in F\}$, $w\in W_0$. Then again by the strong Markov property, we have

$Q_x[w(\tau^x)=y] = (1-Q_x[w(\sigma)=x])^{-1}Q[w(\sigma)=y]$ for $x,y\in F$ with $x\neq y$.

Step 2. Now let $x,y,y' \in F$ and assume that $0 < |x-y| < |x-y'|$. Let

$\xi = \{(s_1,\dots,s_m);\ s_1,\dots,s_m$ is a 1-walk such that $s_1=x$, $s_m=y$, and $s_i\in F^{(1)}\setminus F$, $i = 2,\dots,m-1\}$,

$U = U_{yy'}$, and

$H = \{ z\in\mathbb{R}^D;\ |z-y|\leqq|z-y'|\}$.

Also, we define a map $T:\mathbb{R}^D\to\mathbb{R}^D$ by

$$Tz = \begin{cases} z & \text{if } z\in H \\ Uz & \text{otherwise} \end{cases}.$$

Then we have the following.

(3.6) $Tz \in F^{(1)}$ if $z \in F^{(1)}$.

(3.7) $|Tz_1-Tz_2| \leqq |z_1-z_2|$ for any $z_1,z_2 \in F^{(1)}$.

(3.8) If $z_1,z_2 \in F^{(1)}$ and $Tz_1\neq Tz_2$, then $\tilde{\rho}(Tz_1,Tz_2) \geqq \tilde{\rho}(z_1,z_2)$.

(3.9) $q(Tz_1,Tz_2) \geqq q(z_1,z_2)$ for any $z_1,z_2\in F^{(1)}$.

In fact, (3.6) and (3.7) are obvious. Let us prove (3.8). If $z_1,z_2\in H$ or $z_1,z_2\in\mathbb{R}^D\setminus H$, then this is obvious. Assume that $z_1\in H$ and $z_2\in\mathbb{R}^D\setminus H$. Then if a cell C contains z_1 and z_2, by the assumption (A-2) of symmetry, we see that $z_1,Uz_2\in C$. So we have the assertion (3.8). The case where $z_1\in\mathbb{R}^D\setminus H$ and $z_2\in H$ is similar.

(3.9) is obvious if $Tz_1=Tz_2$. In the case when $Tz_1\neq Tz_2$, since

$\rho(Tz_1)=\rho(z_1)$, the assertion (3.9) follows from (3.7) and (3.8).

Step 3. Let $\xi_0 = \{(Ts_1,\dots,Ts_m);\ (s_1,\dots,s_m)\in\xi\}$. For each $t=(t_1,\dots,t_m)\in\xi_0$, let

$$I(t) = \{i\in\{1,\dots,m-1\};\ t_{i+1}\neq Ut_{i+1},\ \text{and } t_i \text{ and } Ut_{i+1} \text{ belongs to the same 1-cell}\}.$$

Also, for each $t=(t_1,\dots,t_m)\in\xi_0$ and $J=\{i_1,\dots,i_\ell\}\subset I(t)$, let

$$s(J;t) = (t_1,\dots,t_{i_1},Ut_{i_1+1},\dots,Ut_{i_2},U^2t_{i_2+1},\dots,U^{\ell-2}t_{i_{\ell-1}},\\ U^{\ell-1}t_{i_{\ell-1}+1},\dots,U^{\ell-1}t_{i_\ell},U^{\ell}t_{i_\ell+1},\dots,U^{\ell}t_m).$$

Then we easily see that if $s(J,t)=(s_1,\dots,s_m)$, then $Ts_k=t_k$, $k=1,\dots,m$, and that $\xi = \{s(J,t);\ t\in\xi_0,\ J\subset I(t),\ \#(J) \text{ is odd}\}$. Also, we see that if $t,t'\in\xi_0$, $J\subset I(t)$, $J'\subset I(t')$ and $(t,J)\neq(t',J')$, then $s(J,t)\neq s(J',t')$.

For each $s=(s_1,\dots,s_m)\in \bigcup_{k=1}^{\infty}(F^{(1)})^k$, let $q(s) = \prod_{i=1}^{m-1} q(s_i,s_{i+1})$.

Also, we let

$$\tilde{q}(j,s) = \begin{cases} \dfrac{q(s_j,Us_{j+1})}{q(Ts_j,Ts_{j+1})} & \text{if } q(Ts_j,Ts_{j+1}) > 0 \\ 0 & \text{if } q(Ts_j,Ts_{j+1}) = 0 \end{cases},\quad j=1,\dots,m-1.$$

Then by (3.9) and the fact that $TUs_{j+1} = Ts_{j+1}$, we see that $\tilde{q}(j,s) \leqq 1$. Note that for each $t=(t_1,\dots,t_m)\in\xi_0$ and $J=\{j_1,\dots,j_\ell\}\subset I(t)$, we have

$$\begin{aligned} &q(s(J,t)) \\ &= q(t_1,t_2)\dots q(t_{j_1-1},t_{j_1})q(t_{j_1},Ut_{j_1+1})q(Ut_{j_1+1},Ut_{j_1+2})\dots \\ &\quad q(Ut_{j_2-1},Ut_{j_2})q(Ut_{j_2},U^2t_{j_2+1})\dots \\ &= q(t)\cdot \prod_{j\in J}\tilde{q}(j,t). \end{aligned}$$

Now we see that

$$\begin{aligned} &Q_x[\ w(\sigma) = y'\ ] = \sum_{s\in\xi} q(s) \\ &= \sum_{t\in\xi_0}\ \sum_{\substack{J\subset I(t)\\ \#(J)\ \text{odd}}} q(s(J,t)), \end{aligned}$$

and

$$Q_x[\ w(\sigma) = y\ ] \geq \sum_{t\in\xi_0} \sum_{\substack{J\subset I(t)\\ \#(J)\ \text{even}}} q(s(J,t)).$$

Therefore we have

$$Q_x[\ w(\sigma)=y\ ] - Q_x[\ w(\sigma)=y'\ ]$$

$$\geq \sum_{t\in\xi_0} \sum_{J\subset I(t)} (-1)^{\#(J)} q(s(J,t))$$

$$= \sum_{t\in\xi_0} q(t)\cdot \prod_{j\in I(t)} (1-\tilde{q}(j,t)) \geq 0.$$

This and the result in Step 1 imply that $p_{xy}(p) \geq p_{xy'}(p)$.

This completes the proof.

(3.10)**Theorem.** *There is a* $p=(p_1,\ldots,p_r)\in\mathcal{P}$ *such that* $\tilde{p}(p) = p$.

Proof. Since $\tilde{p}:\mathcal{P}\to\mathcal{P}$ is continuous and $\mathcal{P}$ is a compact convex set, we have our assertion by Brouwer's fixed point theorem.

From now on, we fix a $p=(p_1,\ldots,p_r)\in\mathcal{P}$ with $\tilde{p}(p)=p$. As we already saw, $p_1 \geq (\#(F)-1)^{-1}>0$. By Proposition(2.9), we see that for any $x,y \in F$, there is a strict 1-walk $s_1,\ldots,s_m$ such that $s_1=x$, $s_m=y$ and $s_k\in F^{(1)}\setminus F$, $k=2,\ldots,m-1$. So we see that $P_x[\ w(\tau_x)=y\ ] \geq p_1{}^{m-1} > 0$. This proves that $p_i > 0$, $i=1,\ldots,r$.

For any $B \subset F^{(1)}$ and $w \in W_1$, let

$\sigma_B(w) = \min\{\ n \geq 0;\ w(n) \in B\ \}$, and

$\tilde{\sigma}_B(w) = \min\{\ n \geq 1;\ w(n) \in B\ \}$.

For any $i\in I$, let $u_i:F^{(1)}\to[0,1]$ be given by

$u_i(x) = P_x[\ w(\sigma_F) = \pi(\langle i\rangle)\ ]$, $x \in F^{(1)}$.

Also we define a linear map $P:C(F^{(1)};\mathbb{R})\to C(F^{(1)};\mathbb{R})$ by

$Pf(x) = E^{P_x}[\ f(w(1))\ ]$, $f\in C(F^{(1)};\mathbb{R})$, $x\in F^{(1)}$.

Then we have the following.

(3.11)**Proposition.** (1) $P_x[\ w(\tilde{\sigma}_F)=x\ ]$, $x\in F$, *is indehendent of* x.

(2) $Pu_i(x) = u_i(x)$, $i \in I$, $x \in F^{(1)} \setminus F$.

(3) $Pu_i(x) = (1-c)E^{P_x}[u_i(w(\tau^x))] + cu_i(x)$, $i \in I$, $x \in F$,

where $c = P_x[\ w(\tilde{\sigma}_F)=x\ ]$, $x \in F$.

Proof. (1) is obvious from the assumption (A-2) of symmetry (see the proof of Proposition(3.3)). Let $x \in F^{(1)}$. It is obvious that $\tilde{\sigma}_F = 1 + \sigma_F \circ \theta_1$. Here $\theta_n : W_1 \to W_1$, $n \geq 0$, is a map given by $\theta_n w(k) = w(n+k)$, $k \geq 0$, as usual. Thus we have by the Markov property,

$$E^{P_x}[\ u_i(w(\tilde{\sigma}_F))\ ] = E^{P_x}[\ E^{P_{w(1)}}[\ u_i(w(\sigma_F))]] = Pu_i(x).$$

Since $P_x[\ \sigma_F = \tilde{\sigma}_F\ ] = 1$, $x \in F^{(1)} \setminus F$, we have our assertion (2).

Let $x \in F$. Then we have

$$Pu_i(x) = E^{P_x}[\ u_i(w(\tilde{\sigma}_F)),\ \tilde{\sigma}_F = \tau^x] + E^{P_x}[\ u_i(w(\tilde{\sigma}_F)),\ \tilde{\sigma}_F \neq \tau^x] .$$

Note that

$$E^{P_x}[\ u_i(w(\tilde{\sigma}_F)),\ \tilde{\sigma}_F \neq \tau_x\ ] = E^{P_x}[\ u_i(w(\tilde{\sigma}_F)),\ w(\tilde{\sigma}_F)=x]$$
$$= P_x[\ w(\tilde{\sigma}_F)=x\ ]\cdot u_i(x)\ ,$$

and

$$E^{P_x}[\ u_i(w(\tilde{\sigma}_F)),\ \tilde{\sigma}_F = \tau^x\ ]$$
$$= E^{P_x}[\ u_i(w(\tau^x))] - E^{P_x}[\ u_i(w(\tau^x)),\ \tilde{\sigma}_F \neq \tau^x\ ]$$
$$= E^{P_x}[\ u_i(w(\tau^x))] - E^{P_x}[\ u_i(w(\tau^x)),\ \tau^x \geq \tilde{\sigma}_F,\ w(\tilde{\sigma}_F)=x\ ]$$
$$= E^{P_x}[\ u_i(w(\tau^x))] - E^{P_x}[\ E^{P_{w(\tilde{\sigma}_F)}}[u_i(w(\tau^x))],\ w(\tilde{\sigma}_F) = x\ ]$$
$$= (1-P_x[w(\tilde{\sigma}_F) = x])E^{P_x}[u_i(w(\tau^x))]\ .$$

Combining them, we have our assertion (3).

This completes the proof.

It is obvious that $c > 0$. Let us define $q : I \times I \to \mathbb{R}$ by

$$q(i,j) = \begin{cases} p_k & \text{if } (\pi(\langle i \rangle),\pi(\langle j \rangle)) \in C_k\ ,\ k=1,\dots,r \\ -1 & \text{if } i=j \end{cases} .$$

Then it is easy to see that

$$\sum_{j \in I} q(i,j) = 0,\ i \in I,$$

$$q(i,j) = q(j,i),\ i,j \in I,$$

and

$$q(i,j) > 0,\ i\neq j,\ i,j\in I.$$

Then we have the following.

(3.12)**Lemma.** *For any* $\ell\in I$ *and* $f \in C(F^{(1)};\mathbb{R})$,

$$\sum_{k=0}^{N} \{ \sum_{i,j\in I} q(i,j)u_\ell(\pi(\langle ki\rangle)f(\pi(\langle kj\rangle)) \} = (1-c)\cdot \sum_{j\in I} q(\ell,j)f(\pi(\langle j\rangle)).$$

In particular,

$$\sum_{k=0}^{N} \{ \sum_{i,j\in I} q(i,j)u_\ell(\pi(\langle ki\rangle)u_{\ell'}(\pi(\langle kj\rangle)) \} = (1-c)q(\ell,\ell')$$

for any $\ell,\ell'\in I$.

Proof. Note that for any $f,g\in C(F^{(1)};\mathbb{R})$,

$$\sum_{x\in F^{(1)}} \rho(x)((Pg)(x)-g(x))f(x)$$

$$= \sum_{\substack{x,y\in F^{(1)}\\ x\neq y}} p_{xy}\cdot\tilde{\rho}(x,y)g(y)f(x) - \sum_{x\in F^{(1)}} \rho(x)g(x)f(x)$$

(Here $p_{xy} = p_k$ if $|x-y| = \alpha^{-1}\ell_k$)

$$= \sum_{k=0}^{N} \{ \sum_{\substack{x,y\in\psi_k(F)\\ x\neq y}} p_{x,y}g(y)f(x) - \sum_{x\in\psi_k(F)} g(x)f(x) \}$$

$$= \sum_{k=0}^{N} \sum_{i,j\in I} q(i,j)g(\pi(\langle ki\rangle))f(\pi(\langle kj\rangle)) .$$

So letting $g = u_\ell$, by Lemma(2.8) and Proposition(3.11), we see that

$$\sum_{k=0}^{N} \sum_{i,j\in I} q(i,j)u_\ell(\pi(\langle ki\rangle))f(\pi(\langle kj\rangle))$$

$$= \sum_{x\in F} (1-c)(E^{P_x}[u_\ell(w(\tau^x))]-u_\ell(x))f(x)$$

$$= (1-c)\cdot\{ \sum_{\substack{x,y\in F\\ x\neq y}} \tilde{p}_{xy}u_\ell(y)f(y) - \sum_{x\in F} u_\ell(x)f(x)\}$$

(Here $\tilde{p}_{xy} = p_k$ if $|x-y| = \ell_k$)

$$= (1-c)\cdot \sum_{i,j\in I} q(i,j)u_\ell(\pi(\langle i\rangle))f(\pi(\langle j\rangle)) .$$

Note that

$$u_\ell(x) = \begin{cases} 1 & \text{if } x = \pi(\langle\ell\rangle) \\ 0 & \text{if } x \in F\setminus\{\pi(\langle\ell\rangle)\} \end{cases} .$$

These imply our assertion.

4. Dirichlet form on nested fractal.

(4.1)**Definition.** *For each* $n \geq 0$ *and* $f\in C(F^{(n)};\mathbb{R})$, *we define* $\tilde{S}_n f \in C(F^{(n+1)};\mathbb{R})$ *by*

$$\tilde{S}_n f(x) = f(x) \text{ if } x \in F^{(n)},$$

and

$$\tilde{S}_n f(x) = \sum_{i\in I} f(\pi(\langle k_1 \dots k_n i\rangle)) u_i(\psi_{k_n}^{-1}(\dots(\psi_{k_1}^{-1}(x))\dots)$$

$$\text{if } x \in \psi_{k_1}(\dots(\psi_{k_n}(F^{(1)}\setminus F))\dots),\ k_1,\dots,k_n=0,\dots,N.$$

(4.2)**Remark.** *Since* $u_i(y) \geq 0$ *and* $u_i(\pi(\langle i\rangle))=1$, $i\in I$, $y\in F^{(1)}$, *and* $\sum_{i\in I} u_i(y) = 1$, $y\in F^{(1)}$, *we see that*

$$\tilde{S}_n f(x) = \sum_{i\in I} f(\pi(\langle k_1 \dots k_n i\rangle)) u_i(\psi_{k_n}^{-1}(\dots(\psi_{k_1}^{-1}(x))\dots)$$

for any $x \in F^{(1)}_{k_1\dots k_n}$, $k_1,\dots,k_n=0,\dots,N$,

$$\max\{\tilde{S}_n f(x);\ x\in F^{(1)}_{k_1\dots k_n}\} = \max\{f(x);\ x\in F_{k_1\dots k_n}\}$$

and

$$\min\{\tilde{S}_n f(x);\ x\in F^{(1)}_{k_1\dots k_n}\} = \min\{f(x);\ x\in F_{k_1\dots k_n}\}$$

for any $k_1,\dots,k_n=0,\dots,N$.

Let $F^{(\infty)}$ denote $\bigcup_{n=1}^{\infty} F^{(n)}$ and $\tilde{\mathcal{D}}$ denote the set of real valued functions on $F^{(\infty)}$.

(4.3)**Definition.** *For each* $n \geq 0$, $S_n:\tilde{\mathcal{D}}\to\tilde{\mathcal{D}}$ *is defined by*

$$S_n f(x) = \lim_{m\to\infty} \tilde{S}_{n+m}\tilde{S}_{n+m-1}\dots\tilde{S}_n(f|_{F^{(n)}})(x),\ f\in\tilde{\mathcal{D}},\ x\in F^{(\infty)}.$$

This is well-defined because $\tilde{S}_m g(x) = g(x)$ *if* $x\in F^{(\ell)}$, $g\in C(F^{(\ell)};\mathbb{R})$ *and* $m \geq \ell$.

(4.4)**Remark.** *By Definition*(4.3) *and Remark*(4.2), *we have*

$$S_{n+1}S_n f = S_n f,\ n \geq 0,\ f\in\tilde{\mathcal{D}},$$

$$\sup\{S_n f(x);\ x\in F^{(\infty)}_{k_1\dots k_n}\} = \max\{f(x);\ x\in F_{k_1\dots k_n}\}$$

and

$$\inf\{S_n f(x);\ x\in F^{(\infty)}_{k_1\ldots k_n}\} = \min\{f(x);\ x\in F_{k_1\ldots k_n}\}$$

for any $k_1,\ldots,k_n=0,\ldots,N$, and $f\in\tilde{\mathcal{D}}$.

(4.5)**Definition.** *For each* $n \geqq 0$, *we define a symmetric bilinear form* $\mathcal{E}^{(n)}:\tilde{\mathcal{D}}\times\tilde{\mathcal{D}}\to\mathbb{R}$ *by*

$$\mathcal{E}^{(n)}(f,g)$$

$$= -(1-c)^{-n}\sum_{k_1\ldots k_n=0}^{N}\ \sum_{i,j\in I} q(i,j)f(\pi(\langle k_1\ldots k_n i\rangle))g(\pi(\langle k_1\ldots k_n j\rangle))$$

$f,g\in\tilde{\mathcal{D}}$.

(4.6)**Lemma.** (1) $\mathcal{E}^{(n)}(f,f) \geqq 0$ *for any* $n \geqq 0$ *and* $f\in\tilde{\mathcal{D}}$.
(2) $\mathcal{E}^{(n+1)}(S_n f,g) = \mathcal{E}^{(n)}(f,g)$ *for any* $n \geqq 0$, $f,g\in\tilde{\mathcal{D}}$.

Proof. (1) is obvious because $\{q(i,j)\}_{i,j\in I}$ is a non-positive symmetric matrix. By Lemma(3.12) and Remark(4.2), we have

$$-\mathcal{E}^{(n+1)}(S_n f,g)$$

$$= (1-c)^{-(n+1)}\sum_{k_1\ldots k_{n+1}=0}^{N}\ \sum_{i,j\in I} q(i,j)(S_n f)(\pi(\langle k_1\ldots k_{n+1} i\rangle))\times g(\pi(\langle k_1\ldots k_{n+1} j\rangle))$$

$$= (1-c)^{-(n+1)}\sum_{k_1\ldots k_n=0}^{N}\ \sum_{k_{n+1}=0}^{N}\ \sum_{i,j\in I} q(i,j)\times\sum_{\ell\in I} f(\pi(\langle k_1\ldots k_n \ell\rangle))u_\ell(\pi(\langle k_{n+1} i\rangle))g(\pi(\langle k_1\ldots k_{n+1} j\rangle))$$

$$= (1-c)^{-(n+1)}\sum_{k_1\ldots k_n=0}^{N}\ \sum_{\ell\in I} f(\pi(\langle k_1\ldots k_n \ell\rangle))\times\sum_{k=0}^{N}\ \sum_{i,j\in I} q(i,j)\, u_\ell(\pi(\langle ki\rangle))g(\psi_{k_1}(\ldots\psi_{k_n}(\pi(\langle kj\rangle))\ldots))$$

$$= (1-c)^{-n}\sum_{k_1\ldots k_n=0}^{N}\ \sum_{\ell\in I} f(\pi(\langle k_1\ldots k_n \ell\rangle))\cdot\sum_{j\in I} q(\ell,j)\, g(\pi(\langle k_1\ldots k_n j\rangle))$$

$$= -\mathcal{E}^{(n)}(f,g).$$

This proves our assertion (2).

(4.7)Proposition. $\mathcal{E}^{(n+1)}(f,f) = \mathcal{E}^{(n)}(f,f) + \mathcal{E}^{(n+1)}(f-S_n f, f-S_n f)$ *for any* $n \geqq 0$ *and* $f \in \tilde{\mathcal{D}}$. *In particular,* $\mathcal{E}^{(n)}(f,f)$ *is non-decreasing in* n.

Proof. By Lemma(4.6), we have

$$\mathcal{E}^{(n+1)}(S_n f, S_n f) = \mathcal{E}^{(n)}(f, S_n f) = \mathcal{E}^{(n)}(f,f),$$

and

$$\mathcal{E}^{(n+1)}(S_n f, f-S_n f) = \mathcal{E}^{(n)}(f, f-S_n f) = 0.$$

So we have

$$\begin{aligned}&\mathcal{E}^{(n+1)}(f,f)\\ &= \mathcal{E}^{(n+1)}(S_n f, S_n f) + 2\cdot\mathcal{E}^{(n+1)}(S_n f, f-S_n f) + \mathcal{E}^{(n+1)}(f-S_n f, f-S_n f)\\ &= \mathcal{E}^{(n)}(f,f) + \mathcal{E}^{(n+1)}(f-S_n f, f-S_n f).\end{aligned}$$

This completes the proof.

(4.8)Proposition. *There is a* $C_0 \in (0,\infty)$ *such that*

$$\max\{|a_i-a_j|;\ i,j\in I\} \leqq C_0\cdot\{-\sum_{i,j\in I} q(i,j)a_i a_j\}^{1/2}$$

for any $(a_i)_{i\in I} \in \mathbb{R}^I$.

Proof. Let $Q = \{q(i,j)\}_{i,j\in I}$. Then, since $q(i,j) > 0$, $i\neq j$, $\exp(tQ)$ is an irreducible symmetric Markov matrix for each $t > 0$. This implies that the Markov process induced by $\{\exp(tQ);\ t\geqq 0\}$ has strong mixing property. This proves our assertion.

(4.9)Proposition. *There is a* $C_1\in(0,\infty)$ *such that*

$$\max\{|f(x)-f(y)|;\ x,y\in F^{(1)}_{k_1\ldots k_n}\} \leqq C_1(1-c)^{n/2}\cdot\mathcal{E}^{(n+1)}(f,f)^{1/2}$$

for any $f \in \tilde{\mathcal{D}}$, $n \geqq 0$, $k_1,\ldots,k_n = 0,1,\ldots,N$.

Proof. By the assumption (A-1) of connectivity, we have

$$\begin{aligned}&\max\{|f(x)-f(y)|;\ x,y\in F^{(1)}_{k_1\ldots k_n}\}\\ &\leqq \sum_{\ell=0}^{N} \max\{|f(x)-f(y)|;\ x,y\in\psi_{k_1}(\ldots(\psi_{k_n}(\psi_\ell(F)))\ldots)\}\\ &\leqq \sum_{\ell=0}^{N} C_0\cdot\{-\sum_{i,j\in I} q(i,j)f(\pi(\langle k_1\ldots k_n\ell i\rangle))f(\pi(\langle k_1\ldots k_n\ell j\rangle))\}^{1/2}\end{aligned}$$

$$\leq C_0(N+1)^{1/2}\cdot\{-\sum_{\ell_1,\dots,\ell_{n+1}=0}\ \sum_{i,j\in I} q(i,j)$$

$$\times f(\pi(\langle\ell_1\dots\ell_{n+1}i\rangle))f(\pi(\langle\ell_1\dots\ell_{n+1}j\rangle))\}^{1/2}$$

$$= C_0(N+1)^{1/2}\cdot(1-c)^{(n+1)/2}\cdot\mathcal{E}^{(n+1)}(f,f).$$

This proves our assertion.

(4.10)**Lemma.** *There is a* $C_2\in(0,\infty)$ *such that*

$$\max\{|f(x)-f(y)|;\ x,y\in F^{(m)}_{k_1\dots k_n}\}\leq C_2(1-c)^{n/2}\cdot\mathcal{E}^{(n+m)}(f,f)^{1/2}$$

for any $f\in\tilde{\mathcal{D}}$, $n\geq 0$, $k_1,\dots,k_n=0,1,\dots,N$, *and* $m\geq 1$.

Proof. Take an $i_0\in I$ and fix it. Let $x_0=\pi(\langle k_1\dots k_n i_0\rangle)\in F^{(1)}_{k_1\dots k_n}$. For any $x\in F^{(m)}_{k_1\dots k_n}$, there is an $\omega\in\Omega$ and $i_1\in I$ such that

$$\pi(\omega)=x,$$

$$\omega(\ell)=k_\ell,\ \ell=1,\dots,n,\ \text{and}$$

$$\omega(\ell)=i_1,\ \ell\geq n+m+1.$$

Let $\omega_t\in\Omega$, $t=0,1,\dots,m$, be given by

$$\omega_t(\ell)=\begin{cases}\omega(\ell), & \ell=1,2,\dots,n+t\\ i_1, & \ell\geq n+t+1\end{cases}.$$

Then $\omega_m=\omega$ and $\pi(\omega_{t-1}),\ \pi(\omega_t)\in F^{(1)}_{\omega(1)\dots\omega(n+t-1)}$, $t=1,\dots,m$. Therefore by Propositions (4.7) and (4.9) we have

$$|f(\pi(\omega_0))-f(x)|$$

$$\leq\sum_{t=1}^{m}|f(\pi(\omega_{t-1}))-f(\pi(\omega_t))|$$

$$\leq C_1\times\sum_{t=1}^{m}(1-c)^{(n+t-1)/2}\cdot\mathcal{E}^{(n+m)}(f,f)^{1/2}$$

$$\leq C_1(1-(1-c)^{1/2})^{-1}(1-c)^{n/2}\cdot\mathcal{E}^{(n+m)}(f,f)^{1/2}$$

Since $\pi(\omega_0)\in F^{(1)}_{k_1\dots k_n}$, we have

$$|f(x_0)-f(\pi(\omega_0))|\leq C_1(1-c)^{n/2}\cdot\mathcal{E}^{(n+m)}(f,f)^{1/2}.$$

These imply that

$$\max\{|f(x)-f(y)|;\ x,y\in F^{(m)}_{k_1\dots k_n}\}$$

$$\leq 2C_1(1-(1-c)^{1/2})^{-1}(1-c)^{n/2}\cdot\mathcal{E}^{(n+m)}(f,f)^{1/2}.$$

This completes the proof.

The following is an easy consequence of Lemma(4.10).

(4.11)**Corollary.**

$$\sup\{|f(x)-f(y)|;\ x,y\in F^{(\infty)}_{k_1\dots k_n}\} \leqq C_2(1-c)^{n/2}(\lim_{m\to\infty}\mathcal{E}^{(m)}(f,f))^{1/2}$$

for any $f\in\tilde{\mathcal{D}}$, $n\geqq 0$, $k_1,\dots,k_n=0,1,\dots,N$.

Let $B=\{f\in\tilde{\mathcal{D}};\ \lim_{n\to\infty}\mathcal{E}^{(n)}(f,f)\leqq 1\}\subset\tilde{\mathcal{D}}$. Then we have the following.

(4.12)**Proposition.** *Suppose that* $\{\omega_n\}_{n=1}^{\infty}\subset\Omega$, $\omega_n\to\omega$ *in* Ω *as* $n\to\infty$ *and* $\pi(\omega_n)\in F^{(\infty)}$. *Then*

$$\lim_{n,m\to\infty}\sup\{|f(\pi(\omega_n))-f(\pi(\omega_m))|;\ f\in B\}=0.$$

Moreover, if $\pi(\omega)\in F^{(\infty)}$, *then*

$$\lim_{n\to\infty}\sup\{|f(\pi(\omega_n))-f(\pi(\omega))|;\ f\in B\}=0.$$

Proof. Since $\omega_n\to\omega$ in Ω as $n\to\infty$, for any $\ell\geqq 1$, we see that $\pi(\omega_n)\in F^{(\infty)}_{\omega(1)\dots\omega(\ell)}$ for sufficiently large n. So we have by Corollary(4.11)

$$\overline{\lim_{n,m\to\infty}}\sup\{|f(\pi(\omega_n))-f(\pi(\omega_m))|;f\in B\}\leqq C_2(1-c)^{\ell/2}.$$

Since ℓ is arbitrary, we have the first assertion.

Suppose that $\omega\in F^{(\infty)}$. Let $\{\omega'_n\}_{n=1}^{\infty}$ be given by $\omega'_{2n}=\omega_n$ and $\omega'_{2n-1}=\omega_\infty$, $n\geqq 1$. Then by the assertion we have proved, we have

$$\lim_{n,m\to\infty}\sup\{|f(\pi(\omega'_n))-f(\pi(\omega'_m))|;\ f\in B\}=0.$$

This implies the second assertion.

(4.13)**Lemma.** $\lim_{\varepsilon\downarrow 0}\sup\{|f(x)-f(y)|;\ f\in B,\ x,y\in F^{(\infty)},\ |x-y|<\varepsilon\}=0.$

In particular, if $f\in\tilde{\mathcal{D}}$ *and* $\lim_{n\to\infty}\mathcal{E}^{(n)}(f,f)<\infty$, *then there is a unique* $\tilde{f}\in C(E;\mathbf{R})$ *such that* $\tilde{f}|_{F^{(\infty)}}=f$.

Proof. Suppose that

$$\overline{\lim_{\varepsilon\downarrow 0}} \sup\{|f(x)-f(y)|;\ f\in B,\ x,y\in F^{(\infty)},\ |x-y|<\varepsilon\} > 0.$$

Then there are $\delta>0$, $f_n\in B$, ω_n, $\omega_n' \in \Omega$, $n=1,2,\ldots$, such that $\pi(\omega_n),\pi(\omega_n')\in F^{(\infty)}$, $|f_n(\pi(\omega_n))-f_n(\pi(\omega_n'))|>\delta$, $n\geq 1$, and $|\pi(\omega_n)-\pi(\omega_n')|\to 0$ as $n\to\infty$. Since $\Omega\times\Omega$ is compact, we may assume that $(\omega_n,\omega_n')\to(\omega,\omega')\in\Omega\times\Omega$ as $n\to\infty$. Since $\pi:\Omega\to\mathbb{R}^D$ is continuous, we see that $\pi(\omega)=\pi(\omega')$.

Case 1. Suppose that $\pi(\omega)\in E\setminus F^{(\infty)}$. Then by Corollary(2.14), $\omega=\omega'$. Let $\tilde{\omega}_n \in\Omega$ be given by $\tilde{\omega}_{2n-1}= \omega_n$ and $\tilde{\omega}_{2n}= \omega_n'$, $n=1,2,\ldots$. Then $\tilde{\omega}_n\to\omega$ as $n\to\infty$. So we have by Proposition(4.2),

$$\lim_{n,m\to\infty} \sup\{|f(\pi(\tilde{\omega}_n))-f(\pi(\tilde{\omega}_m))|;\ f\in B\} = 0.$$

This contradicts our assumption.

Case 2. Suppose that $\pi(\omega)\in F^{(\infty)}$. Then by Proposition(4.2),

$$\lim_{n\to\infty} \sup\{|f(\pi(\omega_n))-f(\pi(\omega))|;\ f\in B\} = 0$$

and

$$\lim_{n\to\infty} \sup\{|f(\pi(\omega_n'))-f(\pi(\omega))|;\ f\in B\} = 0.$$

This contradicts our assumption. Therefore we have the first assertion.

The second assertion is obvious from the first assertion and the fact that $F^{(\infty)}$ is dense in E.

(4.14)**Theorem.** *Let* $\mathcal{D}om(\mathcal{E}) = \{f\in C(E;\mathbb{R});\ \sup_{n\geq 1} \mathcal{E}^{(n)}(f|_{F^{(\infty)}},f|_{F^{(\infty)}})<\infty\}$, *and define a bilinear form* $\mathcal{E}:\mathcal{D}om(\mathcal{E})\times\mathcal{D}om(\mathcal{E})\to\mathbb{R}$ *by*

$$\mathcal{E}(f,g) = \lim_{n\to\infty} \mathcal{E}^{(n)}(f|_{F^{(\infty)}},g|_{F^{(\infty)}}),\quad f,g \in \mathcal{D}om(\mathcal{E}).$$

Then $\mathcal{D}om(\mathcal{E})$ *is a vector space, the bilinear form* $\mathcal{E}$ *is well-defined and they satisfy the following.*

(1) $\mathcal{D}om(\mathcal{E})$ *is dense in* $C(E;\mathbb{R})$.

(2) $\mathcal{D}om(\mathcal{E})$ *is a local Dirichlet form on* $L^2(E,d\nu\circ\pi^{-1})$.

(3) *There is a* $C \in (0,\infty)$ *such that*

$$\max\{|f(x)-f(y)|\} \leqq C\cdot\mathcal{E}(f,f)^{1/2}$$

for any $f \in \mathcal{D}om(\mathcal{E})$.

(4) *The set* $\{f\in\mathcal{D}om(\mathcal{E});\ \mathcal{E}(f,f)+\|f\|^2_{L^2(E,d\nu\circ\pi^{-1})} \leqq 1\}$ *is compact in* $C(E;\mathbb{R})$.

Here remind that ν *is a probability measure on* Ω *given by*

$$\nu(\omega(i)=k_i,\ i=1,\dots,n) = (N+1)^{-n},\ n\geqq 1 \text{ and } k_1,\dots,k_n=0,\dots,N.$$

Proof. Since $\mathcal{E}^{(n)}(h,h)$, $h\in\tilde{\mathcal{D}}$, is non-decreasing,

$$\mathcal{E}^{(n)}(h_1+h_2,h_1+h_2) \leqq 2(\mathcal{E}^{(n)}(h_1,h_1)+\mathcal{E}^{(n)}(h_2,h_2)),$$

and

$$\mathcal{E}^{(n)}(h_1,h_2) = \frac{1}{4}(\mathcal{E}^{(n)}(h_1+h_2,h_1+h_2)-\mathcal{E}^{(n)}(h_1-h_2,h_1-h_2))$$

for any $n \geqq 0$ and $h_1,h_2\in\tilde{\mathcal{D}}$, it is easy to see that $\mathcal{D}om(\mathcal{E})$ is a vector space and the bilinear form $\mathcal{E}$ is well-defined.

The assertion (3) follows from Corollary(4.11), and the assertion (4) follows from Lemma(4.13). So let us prove the assertions (1) and (2).

(1) Assume that $f\in C(E;\mathbb{R})$. Let $\tilde{f}_n= S_n(f|_{F^{(\infty)}}) \in \tilde{\mathcal{D}}$, $n \geqq 0$. Then we see by Lemma(4.6) that

$$\mathcal{E}^{(m)}(\tilde{f}_n,\tilde{f}_n) = \mathcal{E}^{(n)}(f|_{F^{(\infty)}},f|_{F^{(\infty)}}),\ m \geqq n.$$

So by Lemma(4.13), there is an $f_n\in C(E;\mathbb{R})$ for each $n \geqq 0$ such that $f_n|_{F^{(\infty)}}= \tilde{f}_n$. Then by Remark(4.4), we see that

$$\sup\{|f(x)-f_n(x)|;\ x\in F^{(\infty)}_{k_1\dots k_n}\}$$
$$\leqq \sup\{|f(x)-f(y)|;\ x,y \in F^{(\infty)}_{k_1\dots k_n}\}$$

for any $k_1,\dots,k_n=0,1,\dots,N$.

So we have

$$\sup\{|f(x)-f_n(x)|;\ x\in E\}$$
$$\leqq \sup\{|f(x)-f(y)|;\ x,y \in E,\ |x-y| \leqq \alpha^{-n}\cdot\text{diameter}(E)\}$$
$$\to 0 \text{ as } n\to\infty.$$

This proves our assertion (1).

(2) Suppose that $\{f_n\}_{n=1}^{\infty} \subset \mathcal{D}om(\mathcal{E})$ and $\mathcal{E}(f_n-f_m,f_n-f_m)+\|f_n-f_m\|^2_{L^2(d\nu\circ\pi^{-1})} \to 0$, as $n,m \to \infty$. Since

$$\max\{|f(x)|;\ x\in E\} \leqq \max\{|f(x)-f(y)|;\ x,y\in E\} + \|f\|_{L^2(d\nu\circ\pi^{-1})},$$

we see from the assertion (3) that

$$\max\{|f_n(x)-f_m(x)|;\ x\in E\} \to 0 \text{ as } n,m \to \infty.$$

So there is an $f\in C(E;\mathbb{R})$ and $f_n(x)\to f(x)$ uniformly in x as $n\to\infty$. Then we have

$$\mathcal{E}^{(\ell)}((f-f_n)|_{F^{(\infty)}},(f-f_n)|_{F^{(\infty)}})$$
$$= \lim_{m\to\infty} \mathcal{E}^{(\ell)}((f_m-f_n)|_{F^{(\infty)}},(f_m-f_n)|_{F^{(\infty)}})$$
$$\leqq \lim_{m\to\infty} \mathcal{E}(f_m-f_n,f_m-f_n)\ ,\quad n,\ell \geqq 0.$$

This implies that $f \in \mathcal{D}om(\mathcal{E})$ and $\mathcal{E}(f-f_n,f-f_n) \to 0$ as $n\to\infty$. This proves the closedness of the bilinear form $\mathcal{E}$.

Let $f \in \mathcal{D}om(\mathcal{E})$ and $\varphi\in C^1(\mathbb{R};\mathbb{R})$ such that $|\varphi'(t)| \leqq 1$, $t\in\mathbb{R}$. Since $\{q(i,j)\}_{i,j\in I}$ is a symmetric Markov generator, we see that

$$\mathcal{E}^{(n)}(\varphi\circ f|_{F^{(\infty)}},\varphi\circ f|_{F^{(\infty)}}) \leqq \mathcal{E}^{(n)}(f,f),\ n \geqq 0.$$

This implies that $\varphi\circ f \in \mathcal{D}om(\mathcal{E})$ and $\mathcal{E}(\varphi\circ f,\varphi\circ f) \leqq \mathcal{E}(f,f)$. This shows that the bilinear form $\mathcal{E}$ is Markov.

Now let $f,g \in \mathcal{D}om(\mathcal{E})$ and that $supp(f)\cap supp(g) = \phi$. Then since $diameter(E_{k_1\ldots k_n}) = \alpha^{-n}\cdot diameter(E)$. We see that there is an $n_0 \geqq 0$ such that if $n \geqq n_0$ and $k_1,\ldots,k_n=0,\ldots,N$, then $E_{k_1\ldots k_n}\cap supp(f)=\phi$ or $E_{k_1\ldots k_n}\cap supp(g)= \phi$. Then we see that

$$\mathcal{E}(f,g) = \lim_{n\to\infty} \mathcal{E}^{(n)}(f,g) = 0.$$

Therefore the bilinear form $\mathcal{E}$ is a local Dirichlet form on E.

This completes the proof.

(4.15)**Remark.** *Let* $\{P_t\}_{t\geqq 0}$ *be the semigroup of contraction operators in* $L^2(E;d\nu\circ\pi^{-1})$ *associated with the Dirichlet form* $(\mathcal{E},\mathcal{D}om(\mathcal{E}))$. *Then* *by Theorem(4.14)(3), we can regard* P_t, $t>0$, *as a bounded linear operator from* $L^2(E;d\nu\circ\pi^{-1})$ *into* $C(E;\mathbb{R})$. *So if we regard* $\{P_t\}_{t\geq 0}$ *as a family of operators in* $C(E;\mathbb{R})$, *it is a Feller semigroup. So there is a Feller diffusion process on* E *associated with the local Dirichlet form* $(\mathcal{E},\mathcal{D}om(\mathcal{E}))$.

(4.16)**Proposition.** *For any* $f,g\in\mathcal{D}om(\mathcal{E})$,

$$2\cdot\mathcal{E}(gf,f) - \mathcal{E}(g,f^2)$$
$$= -2\cdot\lim_{n\to\infty}(1-c)^{-n}\sum_{k_1\ldots k_n=0}^{N} g(\pi(\langle k_1\ldots k_n\rangle))$$
$$\times \sum_{i,j\in I} q(i,j)f(\pi(\langle k_1\ldots k_n i\rangle))f(\pi(\langle k_1\ldots k_n j\rangle)).$$

Proof. Note that $q(i,j) > 0$, $i\neq j$, $q(i,j) = q(j,i)$, $i,j\in I$, and $\sum_{i\in I} q(i,j) = 0$. Therefore we have

$$\sum_{i,j\in I} q(i,j)b_i a_j^{\,2}$$
$$= \sum_{i,j\in I} q(i,j)b_i(a_j^{\,2}+a_i^{\,2})$$
$$= \sum_{i,j\in I} q(i,j)b_i(a_j-a_i)^2 + 2\cdot\sum_{i,j\in I} q(i,j)b_i a_i a_j .$$

So we have

$$|(-2\cdot\sum_{i,j\in I} q(i,j)b_i a_i a_j + \sum_{i,j\in I} q(i,j)b_i a_j^{\,2}) + 2c\cdot\sum_{i,j\in I} q(i,j)a_i a_j|$$
$$\leqq \sum_{i,j\in I} q(i,j)|b_i-c|(a_i-a_j)^2$$
$$\leqq -2\cdot\max\{|b_i-c|;i\in I\}\cdot\sum_{i,j\in I} q(i,j)a_i a_j$$

for any $\{a_i\}_{i\in I}$, $\{b_i\}_{i\in I}\in\mathbb{R}^I$ and $c\in\mathbb{R}$.

Now by Theorem(4.14), we have

$$2\cdot\mathcal{E}(gf,f) - \mathcal{E}(g,f^2)$$
$$= \lim_{n\to\infty}(1-c)^{-n}\sum_{k_1\ldots k_n=0}^{N}$$
$$\{-2\cdot\sum_{i,j\in I} q(i,j)g(\pi(\langle k_1\ldots k_n i\rangle))f(\pi(\langle k_1\ldots k_n i\rangle))f(\pi(\langle k_1\ldots k_n j\rangle))$$
$$+\sum_{i,j\in I} q(i,j)g(\pi(\langle k_1\ldots k_n i\rangle))f(\pi(\langle k_1\ldots k_n j\rangle))^2\}.$$

Therefore we have

$$\overline{\lim_{n\to\infty}} \Big| 2\cdot\mathcal{E}(gf,f) - \mathcal{E}(g,f^2) + 2\cdot(1-c)^{-n} \sum_{k_1\ldots k_n=0}^{N} g(\pi(\langle k_1\ldots k_n\rangle)) \times \sum_{i,j\in I} q(i,j) f(\pi(\langle k_1\ldots k_n i\rangle)) f(\pi(\langle k_1\ldots k_n j\rangle)) \Big|$$

$$\leq \overline{\lim_{n\to\infty}} \Big(-2\cdot(1-c)^{-n} \sum_{k_1\ldots k_n=0}^{N} \max_{i\in I} |g(\pi(\langle k_1\ldots k_n i\rangle)) - g(\pi\langle k_1\ldots k_n\rangle))| \times \sum_{i,j\in I} q(i,j) f(\pi(\langle k_1\ldots k_n i\rangle)) f(\pi(\langle k_1\ldots k_n j\rangle)) \Big)$$

$$\leq \overline{\lim_{n\to\infty}} \max\{\sup\{|g(x)-g(y)|; x,y\in E_{k_1\ldots k_n}\}; k_1,\ldots,k_n=0,\ldots,N\}\cdot\mathcal{E}^{(n)}(f,f)$$

$$= 0.$$

This proves our assertion.

5. Probability measure induced by random matrices.

Let V_0 be a finite dimensional real vector space with an inner product $(\, , \,)_{V_0}$. Let Y_i, $i=0,\ldots,N$, be a linear operator in V_0 and w_i, $i=0,\ldots,N$, be positive numbers with $\sum_{k=0}^{N} w_i = 1$.

We assume that there are a positive number λ and a strictly positive symmetric operator Q_0 in V_0 such that

$$(5.1) \quad \sum_{k=0}^{N} w_i \, {}^tY_i Q_0 Y_i = \lambda Q_0 .$$

Let $\mathcal{Q} = \{Q;\ Q$ is a non-negative definite symmetric operators in V_0, trace $QQ_0 = 1\}$, and $\mathcal{Q}_+ = \{Q\in\mathcal{Q};\ Q$ is strictly positive definite$\}$.

(5.2)**Definition.** *For each* $Q\in\mathcal{Q}$, *we define a probability measure* $\mu^{(Q)}$ *on* Ω *by*

$$\mu^{(Q)}(\ \omega(1)=i_1,\ldots,\ \omega(n)=i_n\ )$$
$$= \lambda^{-n} w_{i_1}\cdots w_{i_n} \cdot \mathrm{trace}(Q\,{}^tY_{i_1}\cdots{}^tY_{i_n} Q_0 Y_{i_n}\cdots Y_{i_1})$$

for any $n \geq 1$ *and* $i_1,\ldots,i_n\in\{0,\ldots,N\}$.

(5.3)**Remark.** *By virtue of* (5.1), *the definition of* $\mu^{(Q)}$ *satisfies the consistency condition. So by Kolmogorov's theorem, there is a unique* $\mu^{(Q)}$ *satisfying Definition*(5.2).

The following is obvious.

(5.4)**Proposition.** (1) *If* $Q \in \mathcal{Q}$ *and* $Q' \in \mathcal{Q}_+$, *then* $\mu^{(Q)}$ *is absolutely continuous relative to* $\mu^{(Q')}$.

(2) *For each* $Q \in \mathcal{Q}$, *let* $Q' = \lambda^{-1}\cdot \sum_{k=0}^{N} Y_k Q_0 \, {}^tY_k$. *Then* $Q'\in\mathcal{Q}$ *and* $\mu^{(Q)}\circ T^{-1} = \mu^{(Q')}$. *In particular,* $\mu^{(Q)}\circ T^{-1}$ *is absolutely continuous relative to* $\mu^{(Q)}$, if $Q \in \mathcal{Q}_+$.

(5.5)**Proposition.** *If* $\sum_{k\neq i} {}^tY_k Y_k$ *is strictly positive for all*

$i=0,\dots,N$, *then* $\mu^{(Q)}$ *is non-atomic for any* $Q \in \mathcal{Q}$.

Proof. BY (5.1), we have

$$\lambda^{-1} w_j\, {}^tY_j Q_0 Y_j = Q_0 - \lambda^{-1} \cdot \sum_{k \neq j} {}^tY_k Q_0 Y_k \ , \ j=0,\dots,N.$$

Since by the assumption, $\sum_{k\neq j} {}^tY_k Q_0 Y_k$ is strictly positive definite for all $j=0,\dots,N$, there is an $\varepsilon > 0$ such that

$$\lambda^{-1} w_j\, {}^tY_j Q_0 Y_j \leqq (1-\varepsilon) Q_0 \ , \ j=0,1,\dots,N.$$

So for any $i_1,\dots,i_n \in \{0,\dots,N\}$,

$$\lambda^{-n} w_{i_1} \cdots w_{i_n}\, {}^tY_{i_1} \cdots {}^tY_{i_n} Q_0 Y_{i_n} \cdots Y_{i_1} \leqq (1-\varepsilon)^{-n} Q_0 \ .$$

Therefore we have $\mu^{(Q)}(\omega(k)=i_k,\ k=1,\dots,n) \leqq (1-\varepsilon)^{-n}$. This implies that $\mu^{(Q)}(\{\omega\}) = 0$ for any $\omega \in \Omega$.

This completes the proof.

Let ν be a probability measure on Ω given by

$$\nu(\omega(1)=i_1,\dots,\omega(n)=i_n) = w_{i_1} \cdots w_{i_n} \ , \ n \geqq 1,\ i_1,\dots,i_n \in \{0,\dots,N\}.$$

Let $\mathcal{F}_n^m$ be a σ-algebra on Ω given by $\mathcal{F}_n^m = \sigma\{\omega(k);\ n \leqq k < m+1\}$, $1 \leqq n \leqq m \leqq \infty$. Let $X_n(\omega) = Y_{\omega(n)}$ and $W_n(\omega) = X_n(\omega) \cdots X_1(\omega)$, $n \geqq 1$, $\omega \in \Omega$.

Now let us take a $Q_1 \in \mathcal{Q}_+$ and fix it. We denote $\mu^{(Q_1)}$ by μ. Let $Z_n(\omega)$

$$= \begin{cases} \mathrm{trace}(Q_1\, {}^tW_n(\omega) Q_0 W_n(\omega))^{-1} \cdot {}^tW_n(\omega) Q_0 W_n(\omega), & \text{if } {}^tW_n(\omega) Q_0 W_n(\omega) \neq 0 \\ 0 & , \text{if } {}^tW_n(\omega) Q_0 W_n(\omega) = 0 \end{cases} .$$

Then we have the following.

(5.6)**Proposition.** (1) $\mu(\ {}^tW_n(\omega) Q_0 W_n(\omega) = 0\ ) = 0$. *In particular,* $\mathrm{trace}(Q_1 Z_n(\omega)) = 1$, μ-*a.e.*ω.

(2) $(Z_n(\omega), \mathcal{F}_1^n\ )$ *is a martingale under* μ. *In particular,* $\lim_{n\to\infty} Z_n(\omega) = Z(\omega)$ *exists for* μ-*a.e.*ω.

(3) $\mu(\ {}^tW_n(\omega) Z(T^n\omega) W_n(\omega) = 0\ ) = 0$, *and*

$$Z(\omega) = \mathrm{trace}(Q_1\, {}^tW_n(\omega) Z(T^n\omega) W_n(\omega))^{-1} \cdot {}^tW_n(\omega) Z(T^n\omega) W_n(\omega) \quad \mu\text{-}a.e.\omega$$

for any $n \geq 0$. *Also, for any* $m \geq n$

$$\mathrm{trace}(Q_1\,{}^tW_n(\omega)Z(T^n\omega)W_n(\omega))^{-1}\cdot Z(T^n\omega)$$
$$= \mathrm{trace}(Q_1\,{}^tW_m(\omega)Z(T^m\omega)W_m(\omega))^{-1}\cdot{}^tW_{m-n}(T^n\omega)Z(T^m\omega)W_{m-n}(T^n\omega)\quad \mu\text{-}a.e.\omega .$$

(4) $\mu(\ \mathrm{rank}\ Z(T^{n+1}\omega) \geq \mathrm{rank}\ Z(T^n\omega),\ n \geq 0\ ) = 1.$

(5) *If there is a* $Q \in \mathcal{Q}_+$ *satisfying* $\sum_{k=0}^{N} w_k Y_k Q\,{}^tY_k = \lambda Q$, *then*

$\mu(\ \mathrm{rank}\ Z(T^{n+1}\omega)= \mathrm{rank}\ Z(T^n\omega),\ n \geq 0\ ) = 1.$

(6) *Suppose that* $\sum_{k=0}^{N} w_k Y_k Q_1\,{}^tY_k = \lambda Q_1$. *Then for any* $n \geq 1$ *and*

$f\in C(\{0,\dots,N\}^n;\mathbb{R})$,

$$E^{\mu}[f(\omega(1),\dots,\omega(n))|\mathcal{F}^{\infty}_{n+1}](\omega)$$
$$= \lambda^{-n}\cdot\int_{\Omega} f(\tilde{\omega}(1),\dots,\tilde{\omega}(n))\cdot \mathrm{trace}(Q_1\,{}^tW_n(\tilde{\omega})Z(T^n\omega)W_n(\tilde{\omega}))\ \nu(d\tilde{\omega})$$

for μ-*a.s.*ω.

Proof. (1) By the definition,

$$\mu(\ {}^tW_nQ_0W_n = 0\ )$$
$$= \int_{\{{}^tW_nQ_0W_n=0\}} \lambda^{-n}\cdot \mathrm{trace}(Q_1\,{}^tW_nQ_0W_n)\ d\nu = 0.$$

This proves the assertion (1).

(2) Let $f:\Omega\to\mathbb{R}$ be a bounded $\mathcal{F}^n_1$-measurable function. Then

$$E^{\mu}[f\cdot Z_{n+1}]$$
$$= \lambda^{-n-1}\cdot E^{\nu}[f\cdot{}^tW_{n+1}Q_0W_{n+1}]$$
$$= \lambda^{-n-1}\cdot E^{\nu}[E^{\nu}[f\cdot{}^tW_n\,{}^tX_{n+1}Q_0X_{n+1}W_n|\mathcal{F}^n_1]]$$
$$= \lambda^{-n-1}\cdot E^{\nu}[f\cdot{}^tW_n(\sum_{k=0}^{N} w_k\,{}^tY_kQ_0Y_k)W_n]$$
$$= E^{\mu}[f\cdot Z_n].$$

This shows that $(Z_n, \mathcal{F}^n_1)$ is a martingale under μ. Since $Z_n(\omega)$ is a non-negative definite symmetric matrix and $\mathrm{trace}(Q_1Z_n(\omega))=1$ μ-a.e.ω, each component of $Z_n(\omega)$, $n=1,2,\dots$, is bounded. So by Doob's theorem, $Z_n(\omega)$ converges as $n\to\infty$ for μ-a.e.ω. This proves our assertion (2).

(3) Note that

$$E^{\mu}[\ \mathrm{trace}(Q_1\, {}^tW_n(\omega)Z_k(T^n\omega)W_n(\omega))^{-1}\ ]$$
$$= E^{\mu}[\ \mathrm{trace}(Q_1\, {}^tW_k(T^n\omega)Q_0W_k(T^n\omega))\mathrm{trace}(Q_1\, {}^tW_{n+k}(\omega)Q_0W_{n+k}(\omega))^{-1}\ ]$$
$$= \lambda^{-n-k}\cdot E^{\nu}[\ \mathrm{trace}(Q_1\, {}^tW_k(T^n\omega)Q_0W_k(T^n\omega))\ ]$$
$$= \lambda^{-n-k}\cdot E^{\nu}[\ \mathrm{trace}(Q_1\, {}^tW_k(\omega)Q_0W_k(\omega))\ ]$$
$$= \lambda^{-n}.$$

So letting $k\to\infty$, by Fatou's lemma, we have

$$E^{\mu}[\ \mathrm{trace}(Q_1\, {}^tW_n(\omega)Z(T^n\omega)W_n(\omega))^{-1}\ ] \leqq \lambda^{-n},\ n \geqq 0.$$

This proves that

$$\mu(\ {}^tW_n(\omega)Z_k(T^n\omega)W_n(\omega) = 0\ ) = 0.$$

Since $Z_{n+k}(\omega) = \mathrm{trace}(Q_1\, {}^tW_n(\omega)Z_k(T^n\omega)W_n(\omega))^{-1}\cdot {}^tW_n(\omega)Z_k(T^n\omega)W_n(\omega)$ μ-a.e.ω, letting $k\to\infty$, we have

$$Z(\omega) = \mathrm{trace}(Q_1\, {}^tW_n(\omega)Z(T^n\omega)W_n(\omega))^{-1}\cdot {}^tW_n(\omega)Z(T^n\omega)W_n(\omega) \quad \mu\text{-a.e.}\omega$$

for any $n \geqq 0$. So we have

$$Z(T^n\omega)$$
$$= \mathrm{trace}(Q_1\, {}^tW_\ell(T^n\omega)Z(T^{n+\ell}\omega)W_\ell(T^n\omega))^{-1}\cdot {}^tW_\ell(T^n\omega)Z(T^{n+\ell}\omega)W_\ell(T^n\omega),$$

and this implies that

$$\mathrm{trace}(Q_1\, {}^tW_n(\omega)Z(T^n\omega)W_n(\omega))^{-1}\cdot Z(T^n\omega)$$
$$= \mathrm{trace}(Q_1\, {}^tW_n(\omega)\, {}^tW_\ell(T^n\omega)Z(T^{n+\ell}\omega)W_\ell(T^n\omega)W_n(\omega))^{-1}$$
$$\times {}^tW_\ell(T^n\omega)Z(T^{n+\ell}\omega)W_\ell(T^n\omega)$$
$$= \mathrm{trace}(Q_1\, {}^tW_{n+\ell}(\omega)Z(T^{n+\ell}\omega)W_{n+\ell}(\omega))^{-1}\cdot {}^tW_\ell(T^n\omega)Z(T^{n+\ell}\omega)W_\ell(T^n\omega).$$

This proves our assertion (3).

The assertion (4) follows from the assertion (3) immediately.

(5) Let Q be as in the assumption. Then by Proposition(5.4), we see that $\mu^{(Q)}$ is stationary, and μ and $\mu^{(Q)}$ are mutually absolutely continuous. Then we see that

$$\int_\Omega (\mathrm{rank}\ Z(T^{n+1}\omega)-\mathrm{rank}\ Z(T^n\omega))\ d\mu^{(Q)} = 0,\ n \geqq 0.$$

Since $\mathrm{rank}(Z(T^{n+1}\omega))-\mathrm{rank}(Z(T^n\omega)) \geqq 0$ $\mu^{(Q)}$-a.e.ω by the assertion (4), we see that $\mathrm{rank}(Z(T^{n+1}\omega)) = \mathrm{rank}(Z(T^n\omega))$ $\mu^{(Q)}$-a.e.ω. This implies our assrtion (5).

The assertion (6) follows from the following.

$$E^{\mu}[f(\omega(1),\dots,\omega(n))|\mathcal{F}_{n+1}^{\infty}]$$
$$= \lim_{m\to\infty} E^{\mu}[f(\omega(1),\dots,\omega(n))|\mathcal{F}_{n+1}^{n+m}]$$
$$= \lim_{m\to\infty} \sum_{k_1,\dots,k_n=0}^{N} \lambda^{-n} f(k_1,\dots,k_n) w_{k_1}\cdots w_{k_n}$$
$$\times \mathrm{trace}(Q_1\,{}^tY_{k_1}\cdots{}^tY_{k_n} Z_m(T^n\omega) Y_{k_n}\cdots Y_{k_1}).$$

This completes the proof.

(5.7)**Proposition.** *Let* $a(p) = \inf_{n\geq 1} \lambda^{-n}\cdot E^{\nu}[\|\overset{p}{\Lambda} W_n\|_{\overset{p}{\Lambda} V_0\to\overset{p}{\Lambda} V_0}^{2/p}]$, $p=1,\dots,\dim V_0$. *Then we have the following.*

(1) $a(p) \geq 1$ *or* $a(p) = 0$.

(2) $E^{\mu}[\|\overset{p}{\Lambda} Z_n\|_{\overset{p}{\Lambda} V_0\to\overset{p}{\Lambda} V_0}^{1/p}] = 0$, *if and only if* $a(p) = 0$.

Proof. (1) Let $a_n(p) = \lambda^{-n}\cdot E^{\nu}[\|\overset{p}{\Lambda} W_n\|_{\overset{p}{\Lambda} V_0\to\overset{p}{\Lambda} V_0}^{2/p}]$. Then we have

$$a_{n+m}(p)$$
$$\leq \lambda^{-n-m}\cdot E^{\nu}[\|\overset{p}{\Lambda} W_n(\omega)\|_{\overset{p}{\Lambda} V_0\to\overset{p}{\Lambda} V_0}^{2/p}\cdot\|\overset{p}{\Lambda} W_m(T^n\omega)\|_{\overset{p}{\Lambda} V_0\to\overset{p}{\Lambda} V_0}^{2/p}]$$
$$= a_n(p)a_m(p).$$

So if $a(p)<1$, then $\lim_{n\to\infty} a_n(p) = 0$. This proves the assertion (1).

(2) Note that

$$E^{\mu}[\|\overset{p}{\Lambda} Z_n\|_{\overset{p}{\Lambda} V_0\to\overset{p}{\Lambda} V_0}^{1/p}] = \lambda^{-n}\cdot E^{\nu}[\|\overset{p}{\Lambda}({}^tW_nQ_0W_n)\|_{\overset{p}{\Lambda} V_0\to\overset{p}{\Lambda} V_0}^{1/p}].$$

Also, observe that

$$\|\overset{p}{\Lambda} Q_0^{-1}\|_{\overset{p}{\Lambda} V_0\to\overset{p}{\Lambda} V_0}^{-1/p}\cdot\|\overset{p}{\Lambda} W_n\|_{\overset{p}{\Lambda} V_0\to\overset{p}{\Lambda} V_0}^{2/p}$$
$$\leq \|\overset{p}{\Lambda}({}^tW_nQ_0W_n)\|_{\overset{p}{\Lambda} V_0\to\overset{p}{\Lambda} V_0}^{1/p}$$
$$\leq \|\overset{p}{\Lambda} Q_0\|_{\overset{p}{\Lambda} V_0\to\overset{p}{\Lambda} V_0}^{1/p}\cdot\|\overset{p}{\Lambda} W_n\|_{\overset{p}{\Lambda} V_0\to\overset{p}{\Lambda} V_0}^{2/p}.$$

Therefore, we see that $E^{\mu}[\|\overset{p}{\Lambda} Z_n\|_{\overset{p}{\Lambda} V_0 \to \overset{p}{\Lambda} V_0}^{1/p}] = 0$, if and only if $\lim_{n\to\infty} a_n(p) = 0$. So we have our assertion (2).

(5.8)**Definition.** *Let* $p_0 = \max\{p \geq 1;\ a(p) \geq 1\}$. *We call this* p_0 *the index of* $(\{Y_0,\ldots,Y_N\},\{w_0,\ldots,w_N\})$.

The following is an easy consequence of Propositions (5.4) and (5.7).

(5.9)**Corollary.** *Let* p_0 *be the index of* $(\{Y_0,\ldots,Y_N\},\{w_0,\ldots,w_N\})$. *Then* $\mu(\text{ rank } Z(T^n\omega) \leq p_0,\ n=0,1,\ldots) = 1$ *and* $\mu(\text{ rank } Z(\omega) = p_0\ ) > 0$.

6. Expression of Dirichlet form.

Now we return to the nested fractals. We think of the situation in Sections 2, 3 and 4. Let $A_k = \{a_{ij}^{(k)}\}_{i,j\in I}$, $k=0,\ldots,N$, be matrices given by

$$a_{ij}^{(k)} = u_j(\psi_k(\pi(\langle i\rangle))),\ i,j \in I,\ k=0,\ldots,N.$$

Then we see that

$$a_{ij}^{(k)} \geqq 0,\ i,j \in I,\ k=0,\ldots,N,$$

$$A_k\mathbb{1} = \mathbb{1},\ \text{i.e.,}\ \sum_{j\in I} a_{ij}^{(k)} = 1,\ k=0,\ldots,N,\ i\in I,$$

and

$$a_{ii}^{(i)} = 1,\ i\in I.$$

Here $\mathbb{1} = \begin{pmatrix} 1 \\ \vdots \\ 1 \end{pmatrix} \in \mathbb{R}^I$.

Let $V_0 = \{x\in (x_j)_{j\in I}\in\mathbb{R}^I;\ \sum_{j\in I} x_j = 0\}$. Then V_0 is a subspace of $\mathbb{R}^I$ of codimension 1. Let $\tilde{Q}_0 = \{-q(i,j)\}_{i,j\in I}$. Then $\tilde{Q}_0$ is a non-negative definite symmetric operaor in $\mathbb{R}^I$, and $\tilde{Q}_0\mathbb{1} = 0$, i.e., $\sum_{j\in I} q(i,j) = 0$, $i\in I$. Let P be the orthogonal projection in $\mathbb{R}^I$ whose image is V_0. Then we have $\tilde{Q}_0 P = \tilde{Q}_0$. Let us define linear operators Q_0, Y_k, $k=0,\ldots,N$, in V_0 by

$$Q_0 = P\tilde{Q}_0|_{V_0},\ \text{and}$$

$$Y_k = PA_k|_{V_0},\ k=0,\ldots,N.$$

Then we see that Q_0 is a strictly positive definite symmetric operator in V_0. By Lemma(3.12), we have

$$\sum_{k=0}^{N} {}^tA_k\tilde{Q}_0A_k = (1-c)\tilde{Q}_0.$$

Therefore we have

$$\sum_{k=0}^{N} \left(\frac{1}{N+1}\right)\cdot {}^tY_kQ_0Y_k = \frac{1-c}{N+1}\cdot Q_0.$$

So letting $w_k = \frac{1}{N+1}$, $k=0,\ldots,N$, and $\lambda = \frac{1-c}{N+1}$, we can apply the results in Section 5. Take a strictly positive definite symmetric operator Q_1 with $\mathrm{trace}(Q_0Q_1) = 1$, and fix it (for example $Q_1 = \frac{1}{N+1}\cdot I_{V_0}$).

The following will be proved in the next section.

(6.1)**Proposition.** $\sum_{j\in I\setminus\{i\}} {}^tY_jY_j$ *is strictly positive for all* $i\in I$.

Then we have the following from Proposition(5.5).

(6.2)**Proposition.** *The probability measure* μ *is non-atomic.*

Now for any $f \in \tilde{\mathcal{D}}$ and $n \geq 0$, let $\tilde{u}_n(f):\Omega\to\mathbb{R}^I$ be given by

$$\tilde{u}_n(f)(\omega) = \{f(\pi(\langle\omega(1)\omega(2)\ldots\omega(n)i\rangle))\}_{i\in I} .$$

Then we have the following.

(6.3)**Lemma.** (1) $\tilde{u}_{n+1}(S_nf)(\omega) = A_{\omega(n+1)}\tilde{u}_n(f)(\omega)$ *for any* $\omega\in\Omega$, $n \geq 0$ *and* $f\in\tilde{\mathcal{D}}$.

(2) *For any* $g\in\mathcal{D}om(\mathcal{E})$, $f\in\tilde{\mathcal{D}}$ *and* $n \geq 1$,

$$- \lim_{m\to\infty} (1-c)^{-m} \sum_{k_1,\ldots,k_m=0}^{N} g(\pi(\langle k_1\ldots k_m\rangle))$$
$$\times \sum_{i,j\in I} q(i,j)(S_nf)(\pi(\langle k_1\ldots k_mi\rangle))(S_nf)(\pi(\langle k_1\ldots k_mj\rangle))$$
$$= E^{\mu}[g(\pi(\omega))\cdot \mathrm{trace}(Q_1{}^tW_n(\omega)Z(T^n\omega)W_n(\omega))^{-1}$$
$$\times(P\tilde{u}_n(f)(\omega),Z(T^n\omega)P\tilde{u}_n(f)(\omega))_{V_0}]$$

In particular,

$$\mathcal{E}^{(n)}(f,f)$$
$$= E^{\mu}[\ \mathrm{trace}(Q_1{}^tW_n(\omega)Z(T^n\omega)W_n(\omega))^{-1}\cdot(P\tilde{u}_n(f)(\omega),Z(T^n\omega)P\tilde{u}_n(f)(\omega))_{V_0}]$$

for any $f\in\tilde{\mathcal{D}}$ and $n\geq 0$.

Proof. The assertion (1) follows from that

$$\tilde{u}_{n+1}(S_nf)_i$$
$$= (S_nf)(\pi(\langle\omega(1)\ldots\omega(n+1)i\rangle))$$
$$= \sum_{j\in I} f(\pi(\langle\omega(1)\ldots\omega(n)j\rangle))u_j(\pi(\langle\omega(n+1)i\rangle)$$
$$= (A_{\omega(n+1)}\tilde{u}_n(f)(\omega))_i .$$

(2) By the assertion (1), we have

$$- \lim_{m\to\infty} (1-c)^{-m} \sum_{k_1,\ldots,k_m=0}^{N} g(\pi(\langle k_1\ldots k_m\rangle))$$

$$\times \sum_{i,j\in I} q(i,j)(S_n f)(\pi(\langle k_1\ldots k_m i\rangle))(S_n f)(\pi(\langle k_1\ldots k_m j\rangle))$$

$$= \lim_{m\to\infty} \lambda^{-(n+m)} E^{\nu}[g(\pi([\omega,\langle\omega(n+m)\rangle]_{n+m}))$$

$$\times (P\tilde{u}_{n+m}(S_n f)(\omega), Q_0 P\tilde{u}_{n+m}(S_n f)(\omega))_{V_0}]$$

$$= \lim_{m\to\infty} \lambda^{-(n+m)} E^{\nu}[g(\pi([\omega,\langle\omega(n+m)\rangle]_{n+m}))$$

$$\times (PA_{\omega(m+n)}\cdots A_{\omega(n+1)}\tilde{u}_n(f)(\omega), Q_0 P\, A_{\omega(m+n)}\cdots A_{\omega(n+1)}\tilde{u}_n(f)(\omega))_{V_0}]$$

$$= \lim_{m\to\infty} \lambda^{-(n+m)} E^{\nu}[g(\pi([\omega,\langle\omega(n+m)\rangle]_{n+m}))$$

$$\times (P\tilde{u}_n(f)(\omega), {}^tW_m(T^n\omega)Q_0 W_m(T^n\omega)P\tilde{u}_n(f)(\omega))_{V_0}]$$

$$= \lim_{m\to\infty} E^{\mu}[\, g(\pi([\omega,\langle\omega(n+m)\rangle]_{n+m}))\cdot \mathrm{trace}(Q_1\, {}^tW_n(\omega)Z_m(T^n\omega)W_n(\omega))^{-1}$$

$$\times (P\tilde{u}_n(f)(\omega), Z_m(T^n\omega)P\tilde{u}_n(f)(\omega))_{V_0}].$$

$$= E^{\mu}[g(\pi(\omega))\cdot \mathrm{trace}(Q_1\, {}^tW_n(\omega)Z(T^n\omega)W_n(\omega))^{-1}$$

$$\times (P\tilde{u}_n(f)(\omega), Z(T^n\omega)P\tilde{u}_n(f)(\omega))_{V_0}]$$

This implies the first half part of the assertion (2). The second half part follows from Lemma(4.6) and the first half part by letting $g=1$.

This completes the proof.

Now let

(6.4) $r(\omega) = \min\{r\geqq 0;\ \mathrm{rank}\ Z(T^k\omega) = \mathrm{rank}\ Z(T^r\omega) \text{ for all } k\geqq r\},\ \omega\in\Omega.$

Then we have $\mu(r(\omega)<\infty) = 1$ by Proposition(5.6). Also, let $V(\omega)$ be the image of $Z(\omega)$ and $P(\omega)$ be the orthogonal projection in V_0 onto $V(\omega)$ for each $\omega\in\Omega$.

(6.5)**Proposition.** (1) *If* $m \geqq n$, *then*

$$\mathrm{trace}(Q_1\, {}^tW_n(\omega)Z(T^n\omega)W_n(\omega))^{-1}(P\tilde{u}_n(f)(\omega), Z(T^n\omega)P\tilde{u}_n(f)(\omega))_{V_0}$$

$$= \mathrm{trace}(Q_1\, {}^tW_m(\omega)Z(T^m\omega)W_m(\omega))^{-1}(P\tilde{u}_m(S_n f)(\omega), Z(T^m\omega)P\tilde{u}_m(S_n f)(\omega))_{V_0}.$$

(2) *If* $r(\omega) = 0$, *then* $P(T^n\omega)W_n(\omega)|_{V(\omega)}: V(\omega)\to V(T^n\omega)$ *is bijective, and*

$P(T^n\omega)W_n(\omega) = P(T^n\omega)W_n(\omega)P(\omega).$

Proof. (1) Note that

$$P\tilde{u}_m(S_n f)(\omega) = PA_{\omega(m)}\cdots A_{\omega(n+1)}\tilde{u}_n(f)(\omega) = W_{m-n}(T^n\omega)P\tilde{u}_n(f)(\omega).$$

Then by Proposition(5.6) we have the assertion (1).

(2) Note that

$$Z(\omega) = \mathrm{trace}(Q_1\,{}^tW_n(\omega)Z(T^n\omega)W_n(\omega))\cdot{}^t(P(T^n\omega)W_n(\omega))Z(T^n\omega)P(T^n\omega)W_n(\omega).$$

and so

$$\dim V(T^n\omega) \geqq \mathrm{rank}\ P(T^n\omega)W_n(\omega) \geqq \mathrm{rank}\ Z(\omega) = \dim P(\omega).$$

Therefore since $r(\omega) = 0$, we have $\mathrm{rank}\ P(T^n\omega)W_n(\omega) = \dim V(\omega)$. If $u \in V(\omega)^{\perp}$, then we have

$$(P(T^n\omega)W_n(\omega)u, Z(T^n\omega)P(T^n\omega)W_n(\omega)u)$$
$$= \mathrm{trace}(Q_1\,{}^tW_n(\omega)Z(T^n\omega)W_n(\omega))\cdot(u, Z(\omega)u)_{V_0} = 0.$$

So $P(T^n\omega)W_n(\omega)u = 0$. This implies that $\mathrm{rank}(P(T^n\omega)W_n(\omega)|_{V(\omega)}) = \dim V(\omega)$. Therefore $P(T^n\omega)W_n(\omega)|_{V(\omega)} : V(\omega) \to V(T^n\omega)$ is bijective.

This completes the proof.

Suppose that $r(\omega) = r$. Then $r(T^r\omega) = 0$. Therefore by Proposition(6.5), we see that $P(T^n\omega)W_{n-r}(T^r\omega)|_{V(T^r\omega)} : V(T^r\omega) \to V(T^n\omega)$, $n \geqq r$, is bijective. So let us define

(6.6) $$\tilde{Z}(\omega) = \mathrm{trace}(Q_1\,{}^tW_{r(\omega)}(\omega)Z(T^{r(\omega)}\omega)W_{r(\omega)}(\omega))^{-1}Z(T^{r(\omega)}\omega)$$

and

(6.7) $$u_n(f)(\omega) = \begin{cases} P(T^{r(\omega)}\omega)\tilde{u}_{r(\omega)}(S_n f;\omega) & \text{if } n \leqq r(\omega) \\ (P(T^n\omega)W_{n-r(\omega)}(T^{r(\omega)}\omega)|_{V(T^{r(\omega)}\omega)})^{-1}P(T^n\omega)\tilde{u}_n(f)(\omega) & \text{if } n > r(\omega) \end{cases}$$

for any $n \geqq 1$, $f \in \tilde{\mathcal{D}}$ and $\omega \in \Omega$.

Then we see that $u_n(f)(\omega) \in V(T^{r(\omega)}\omega)$ and $u_n(f)(\omega)$ is linear in f.

Also we have the following.

(6.8)**Lemma.** (1) $(u_n(f)(\omega), \tilde{Z}(\omega)u_n(f)(\omega))_{V_0}$

$$= \mathrm{trace}(Q_1\,{}^tW_n(\omega)Z(T^n\omega)W_n(\omega))^{-1}(P\tilde{u}_n(f)(\omega), Z(T^n\omega)P\tilde{u}_n(f)(\omega))_{V_0}$$

for any $n \geqq 1$, $f \in \tilde{\mathcal{D}}$ *and* $\omega \in \Omega$.

(2) $u_{n+1}(S_n f)(\omega) = u_n(f)(\omega)$

for any $n \geq 1$, $f \in \tilde{\mathcal{D}}$ *and* $\omega \in \Omega$.

(3) *For any* $g \in \mathcal{D}om(\mathcal{E})$, $f \in \tilde{\mathcal{D}}$ *and* $n \geq 1$,

$$- \lim_{m\to\infty} (1-c)^{-m} \sum_{k_1,\dots,k_m=0}^{N} g(\pi(\langle k_1 \dots k_m\rangle))$$

$$\times \sum_{i,j\in I} q(i,j)(S_n f)(\pi(\langle k_1\dots k_m i\rangle))(S_n f)(\pi(\langle k_1\dots k_m j\rangle))$$

$$= E^{\mu}[g(\pi(\omega))\cdot(u_n(f)(\omega),\tilde{Z}(\omega)u_n(f)(\omega))_{V_0}].$$

(4) $\mathcal{E}^{(n)}(f,f) = E^{\mu}[(u_n(f)(\omega),\tilde{Z}(\omega)u_n(f)(\omega))_{V_0}]$, $n \geq 1$,

and

$$\mathcal{E}^{(m)}(f,f) - \mathcal{E}^{(n)}(f,f)$$

$$= E^{\mu}[(u_m(f)(\omega)-u_n(f)(\omega),\tilde{Z}(\omega)(u_m(f)(\omega)-u_n(f)(\omega)))_{V_0}],\ m \geq n \geq 1,$$

for any $f \in \tilde{\mathcal{D}}$.

Proof. The assertion (1) is obvious from Proposition(5.6)(3) and Proposition(6.5). The assertion (2) follows from Proposition(6.3). The assertion (3) follows from the assertion (1) and Lemma(6.3). The assertion (4) follows from Lemma(6.3), the assertions (1), (2) and the fact that

$$\mathcal{E}^{(m)}(f,f) - \mathcal{E}^{(n)}(f,f) = \mathcal{E}^{(m)}(f-S_n f, f-S_n f).$$

This completes the proof.

(6.9)**Proposition.** *For each* $f \in \mathcal{D}om(\mathcal{E})$, *there is a measurable map* $u(f):\Omega \to V_0$ *satisfying the following.*

(1) $E^{\mu}[(u(f)(\omega),\tilde{Z}(\omega)u(f)(\omega))_{V_0}] < \infty$, $f \in \mathcal{D}om(\mathcal{E})$,

(2) $\mu(\ u(f)(\omega) \in \mathrm{Image}\ \tilde{Z}(\omega)\ ) = 1$, $f \in \mathcal{D}om(\mathcal{E})$,

(3) *for any* $f \in \mathcal{D}om(\mathcal{E})$,

$$\lim_{n\to\infty} E^{\mu}[((u(f)(\omega)-u_n(f|_{F^{(\infty)}})(\omega)),\tilde{Z}(\omega)(u(f)(\omega)-u_n(f|_{F^{(\infty)}})(\omega)))_{V_0}] = 0$$

and

(4) $u(af)(\omega)+u(bg)(\omega) = u(af+bg)(\omega)$ μ-*a.e.*ω

for any $f,g \in \mathcal{D}om(\mathcal{E})$ *and* $a,b \in \mathbf{R}$.

Proof. From the definition, we see that $u_n(f|_{F^{(\infty)}})(\omega)\in \mathrm{Image}\ \tilde{Z}(\omega)$ for any $f\in\tilde{\mathcal{D}}$ and $n=0,1,\ldots$. By Lemma(6.8), we see that for any $f\in\mathcal{D}om(\mathcal{E})$

$$\lim_{n,m\to\infty} E^{\mu}[(u_n(f|_{F^{(\infty)}})(\omega)-u_m(f|_{F^{(\infty)}})(\omega),\ \tilde{Z}(\omega)(u_n(f|_{F^{(\infty)}})(\omega)-u_m(f|_{F^{(\infty)}})(\omega)))_{V_0}] = 0.$$

Therefore $(u_n(f|_{F^{(\infty)}})(\omega))_{n=1}^{\infty}$ converges in probability. Letting $u(f)(\omega)$ to be the limit of $(u_n(f|_{F^{(\infty)}})(\omega))_{n=1}^{\infty}$, we have our assertion.

Remember that from the general theory of Dirichlet form (see Fukushima[7, Chapter 5]), for each $f\in\mathcal{D}om(\mathcal{E})$, there is a martingale additive functional $M_t^{[f]}$ associated with f. Moreover, for any $f,g\in\mathcal{D}om(\mathcal{E})$, there is a signed measure $\mu^{[f,g]}$ on E associated with the additive functional $\langle M^{[f]},M^{[g]}\rangle_t$.

(6.10)**Proposition.** *For any* $f,g\in\mathcal{D}om(\mathcal{E})$, $\mu^{[f,g]}(F^{(\infty)})=0$, *and*

$$\mu^{[f,g]} = ((u(f)(\omega),\tilde{Z}(\omega)u(g)(\omega))\mu(d\omega))_{V_0}\circ\pi^{-1}.$$

Proof. By Proposition(4.16) and Lemma(6.8)(3), we have

$$2\cdot\int_E g(x)\ \mu^{[f,f]}(dx)$$
$$= 2\mathcal{E}(gf,f)-\mathcal{E}(g,f^2)$$
$$= 2\cdot E^{\mu}[g(\pi(\omega))(u(f)(\omega),\tilde{Z}(\omega)u(f)(\omega))_{V_0}]$$

for any $f,g\in\mathcal{D}om(\mathcal{E})$. Therefore for any $f\in\mathcal{D}om(\mathcal{E})$ and bounded measurable function $h:E\to\mathbb{R}$,

$$\int_E h(x)\ \mu^{[f,f]}(dx) = E^{\mu}[h(\pi(\omega))(u(f)(\omega),\tilde{Z}(\omega)u(f)(\omega))_{V_0}].$$

Since μ is non-atomic, $\pi^{-1}(F^{(\infty)})$ is countable, and $\pi^{-1}:E\setminus F^{(\infty)}\to\Omega$ is one to one, we see that $\mu^{[f,f]}(F^{(\infty)}) = 0$ and

$$\mu^{[f,f]} = ((u(f)(\omega),\tilde{Z}(\omega)u(f)(\omega))_{V_0}\mu(d\omega))\circ\pi^{-1}.$$

This implies our assertion.

Let $L^2(\tilde{Z})$ denotes the Hilbert space with an inner product $(\ ,\ )_{\tilde{Z}}$ given by

$$L^2(\tilde{Z}) = \{k:\Omega\to V_0;\ k(\omega)\in \text{Image } \tilde{Z}(\omega)\ \mu\text{-a.s.}\omega,\ E^{\mu}[(k(\omega),\tilde{Z}(\omega)k(\omega))_{V_0}]<\infty\},$$

and

$$(k_1,k_2)_{\tilde{Z}} = E^{\mu}[(k_1(\omega),\tilde{Z}(\omega)k_2(\omega))_{V_0}],\quad k_1,k_2\in L^2(\tilde{Z}).$$

Then we have the following.

(6.11)**Proposition.** (1) *If* $f\in\mathcal{D}om(\mathcal{E})$ *and* $h:E\to\mathbb{R}$ *is a bounded measurable function,* $(h\circ\pi)\cdot u(f)\in L^2(\tilde{Z})$.

(2) *Let* $f,g\in\mathcal{D}om(\mathcal{E})$ *and* h_1 *and* h_2 *be bounded functions on* E. *Then*

$$E^{P_m}[\langle h_1M^{[f]},h_2M^{[g]}\rangle_1] = 2\cdot((h_1\circ\pi)u(f),(h_2\circ\pi)u(g))_{\tilde{Z}}\ .$$

Here P_m *is the measure of the associated diffusion process with the stationary initial distribution* $m = \nu\circ\pi^{-1}$.

(3) *The linear span of* $\{(h\circ\pi)u(f);\ f\in\mathcal{D}om(\mathcal{E}),\ h$ *is a bounded function on* E$\}$ *is dense in* $L^2(\tilde{Z})$.

(4) *There are* $k_1,\dots,k_N\in L^2(\tilde{Z})$ *such that*

$$(k_i(\omega),\tilde{Z}(\omega)k_j(\omega))_{V_0} = 0\quad \mu\text{-a.e.}\omega,\ i\neq j,$$

$$\sum_{i=1}^{d(\omega)} \mathbb{R}k_i(\omega) = \mathcal{I}\text{mage } \tilde{Z}(\omega),\ \omega\in\Omega,$$

and

$$(k_i(\omega),\ \tilde{Z}(\omega)k_i(\omega))_{V_0} = 1,\ \textit{if}\ i \leqq d(\omega).$$

Here $d(\omega) = \textit{rank}\ \tilde{Z}(\omega)$, $\omega\in\Omega$.

(5) *There is a natural isometric linear map* ι from $L^2(\tilde{Z})$ *into the space of square integrable martingale additive functionals satisfying*

$$\iota((h\circ\pi)u(f)) = \int_0^{\cdot} h(X_t)dM_t^{[f]}\ .$$

Moreover, the associated signed measure of the additive functional $\langle\iota(k_1),\iota(k_2)\rangle_t$, $k_1,k_2\in L^2(\tilde{Z})$, *is* $((k_1(\omega),\tilde{Z}(\omega)k_2(\omega))_{V_0}\mu(d\omega))\circ\pi^{-1}$.

Proof. The assertions (1) and (2) are obvious. The assertion (5) follows from the assertions (1), (2) and (3). So we shall prove the assertions (3) and (4).

Let $(f_k)_{k=1}^{\infty}$ be a dense subset of $\mathcal{D}om(\mathcal{E})$. Let $f_{k,n}$ be the continuous extension of $S_n(f_k|_{F^{(\infty)}})$. Then $f_{k,n}\in\mathcal{D}om(\mathcal{E})$. Also, by Lemma(6.8), we see that $u_m(f_{k,n}|_{F^{(\infty)}})(\omega) = u_n(f_{k,n}|_{F^{(\infty)}})(\omega)$, $m\geqq n$. Therefore $u(f_{k,n})(\omega) = u_n(f_{k,n}|_{F^{(\infty)}})(\omega)$. It is easy to see that if $n \geqq r(\omega)$, then $\dim(\sum_{k=1}^{\infty} \mathbb{R}u_n(f_{k,n}|_{F^{(\infty)}})(\omega)) = \operatorname{rank}\tilde{Z}(\omega)$. So we see that μ-a.e.ω

$$\dim\Big(\sum_{n=1}^{\infty} \sum_{k=1}^{\infty} \mathbb{R}u(f_{k,n})(\omega) \Big) = \operatorname{rank}\tilde{Z}(\omega).$$

Let $R:V_0\to V_0$ be given by

$$R(v) = \begin{cases} (v,\tilde{Z}(\omega)v)_{V_0}^{-1/2}\, v & \text{if } \tilde{Z}(\omega)v\neq 0 \\ 0 & \text{if } \tilde{Z}(\omega)v=0 \end{cases}.$$

Now let $j:\mathbb{N}\times\mathbb{N}\to\mathbb{N}$ be the one to one onto map and for each $i\in\mathbb{N}$, let $g_i = f_{k,n}$, $(k,n) = j(i)$. Now let us define $e_n(\omega)$, $n=1,2,\ldots$, by

$$e_1(\omega) = R(u(g_1)(\omega)),$$

and

$$e_{n+1}(\omega) = R\Big(u(g_{n+1})(\omega) - \sum_{k=1}^{n} (u(g_{n+1})(\omega),\tilde{Z}(\omega)e_k(\omega))_{V_0} e_k(\omega) \Big)$$

for $n=1,2,\ldots$.

Let $\Omega_j = \{\omega\in\Omega;\ \operatorname{rank}\tilde{Z}(\omega)=j\}$, $j=1,\ldots,N$. For each $j=1,\ldots,N$, let $S_j = \{(i_1,\ldots,i_j)\in\mathbb{N}^j;\ i_1<\ldots<i_j\}$. Moreover, for $s=(i_1,\ldots,i_j)\in \bigcup_{j=1}^{N} S_j$, let $\Omega_s = \{\omega\in\Omega_j;\ e_{i_k}(\omega)\neq 0,\ k=1,\ldots,j\}$. Then it is easy to see that Ω_s, $s\in \bigcup_{j=1}^{N} S_j$, are disjoint and their union is Ω. Let $k_m(\omega)$, $m=1,\ldots,N$, be given by

$$k_m(\omega) = \sum_{j=m}^{N} \sum_{s\in S_m} \chi_{\Omega_s}(\omega)\cdot e_{i_m}(\omega).$$

ere $s = (i_1,\ldots,i_j)$ for $s \in S_j$. Then it is easy to see that $k_1,\ldots,k_N$ satisfies the assertion (4) and that k_m, $m=1,\ldots,N$, belongs to the closure of the linear span of $\{(h\circ\pi)u(f);f\in\mathcal{D}om(\mathcal{E})$, h is a bounded function in E} in $L^2(\tilde{Z})$. This proves our assertions (3) and (4).

This completes the proof.

(6.12)**Theorem.** The martingale dimension of the associated diffusion process is the index p_0 of $\{(Y_0,\dots,Y_N\},\{w_0,w_1,\dots,w_N\}\}$.

Proof. By Proposition(5.4) and Corollary(5.9), we see that $\mu($ rank $\tilde{Z}(\omega) \leqq p_0) = 1$ and $\mu($ rank $\tilde{Z}(\omega) = p_0) > 0$. Let $\Omega_k = \{\omega\in\Omega;$ rank $\tilde{Z}(\omega) \geqq k\}$, $k=1,\dots,p_0$. Let $k_1,\dots,k_N \in L^2(\tilde{Z})$ be as in Proposition(6.11)(3). Then we see that $k_j(\omega) = 0$ μ-a.e.ω, $j > p_0$, and $(k_j(\omega),\tilde{Z}(\omega)k_j(\omega))_{V_0} = \chi_{\Omega_j}(\omega)$ μ-a.e.ω, $j \leqq p_0$. Let $M^j = \iota(k_j)$, $j=1,\dots,p_0$. Then by Proposition(6.11)(5), we see that $\langle M^i,M^j\rangle = 0$, $i\neq j$, and $d\langle M^1,M^1\rangle_t \geqq d\langle M^2,M^2\rangle_t \geqq \dots \geqq d\langle M^{p_0},M^{p_0}\rangle_t$ P_m-a.s. Moreovere, Let M be a square integrable martingale additive functional. Then there are $h_i:\Omega\to\mathbb{R}$, $i=1,\dots,p_0$ such that

$$\iota^{-1}(M)(\omega) = \sum_{j=1}^{p_0} h_j(\omega)k_j(\omega) \quad \mu\text{-a.e.}\omega.$$

Then we have $\displaystyle\sum_{j=1}^{p_0}\int_{\Omega_j} |h_j(\omega)|^2 \mu(d\omega) < \infty$. Let $\tilde{h}_j:E\to\mathbb{R}$ be given by

$$\tilde{h}_j(x) = \begin{cases} h_j(\pi^{-1}(x)) & \text{if } x\in E\setminus F^{(\infty)} \\ 0 & \text{if } x\in F^{(\infty)} \end{cases}.$$

Then we see that $\iota(h_jk_j) = \int_0^{\cdot} \tilde{h}_j(X_t)dM_t^j$, $j=1,\dots,p_0$. Therefore

$$M_t = \sum_{j=1}^{p_0}\int_0^t \tilde{h}_j(X_t)dM_t^j \quad P_m\text{-a.s.}$$

This proves that $(M^j)_{j=1}^{p_0}$ is a martingale basis. So the martingale dimension is p_0.

7. Some remarks for the measure μ.

We use the same notation as in Section 6.

Let $i \in I$. Then we have

$a^{(i)}_{ik} = 0$, $k \in I\setminus\{i\}$,

$a^{(i)}_{ii} = 1$, and

$a^{(i)}_{jk} > 0$, $j \in I\setminus\{i\}$, $k\in I$.

Let $J_i = I\setminus\{i\}$, and $B_i = \{a^{(i)}_{jk}\}_{j,k\in J_i}$. Then we have

$$\det(sI_0 - A) = \pm(s-1)\cdot\det(sI_i - B_i).$$

Here I_0 and I_i are the identity matrix in $\mathbb{R}^I$ and $\mathbb{R}^{J_i}$ respectively. Since $\sum_{k\in J} a^{(i)}_{jk} < 1$ and B_i is a positive matrix, the maximum eigen value s_i of B_i is in $(0,1)$. Moreover, 1 and s_i are simple roots of $\det(sI_0 - A_i) = 0$ and the absolute values of other roots are less than s_i. By Perron-Frobenius's theorem, there are positive vectors $\tilde{u}_i, \tilde{v}_i \in \mathbb{R}^{J_i}$ such that ${}^tB_i\tilde{u}_i = s_i\tilde{u}_i$, $B_i\tilde{v}_i = s_i\tilde{v}_i$, $\|\tilde{u}_i\|_{\mathbb{R}^{J_i}} = 1$ and $(\tilde{u}_i, \tilde{v}_i)_{\mathbb{R}^{J_i}} = 1$.

Let u_i, $i\in I$, be elements of V_0 given by $u_{i,k} = \tilde{u}_{i,k}$, $k\in J_i$ and $u_{i,i} = -\sum_{j\in J_i} \tilde{u}_{i,j}$. Also, let v_i, $i\in I$, be element of V_0 given by $v_{i,k} = \tilde{v}_{i,k}$, $k\in J_i$ and $v_{i,i} = 0$. Then we have ${}^tA_iu_i = s_iu_i$, $A_iv_i = v_i$ and $(u_i, v_i)_{\mathbb{R}^I} = 1$. Since $A_i\mathbb{1} = \mathbb{1}$ and $P\mathbb{1} = 0$, we have

$$\lim_{n\to\infty} s_i^{-n}\cdot PA_i^n x = (u_i, x)_{\mathbb{R}^I}\cdot Pv_i, \quad x\in\mathbb{R}^I.$$

Since $PA_i = PA_iP$ and $u_i\in V_0$, we have

$$\lim_{n\to\infty} s_i^{-n}\cdot Y_i^n x = (u_i, x)_{V_0}\cdot Pv_i, \quad x\in\mathbb{R}^I.$$

If $x = (x_i)_{i\in I}\in\mathbb{R}^I$ and $|x_\ell| = \max(|x_i|;\ i\in I)$, $\ell\in I$, we have

$$|(x, u_\ell)_{V_0}| = \Big|\sum_{j\in I} x_j u_{\ell,j}\Big| = \Big|x_\ell\Big(\sum_{j\in J_\ell}\tilde{u}_{\ell,j}\Big) - \sum_{j\in J_\ell} x_j\tilde{u}_{\ell,j}\Big|$$

$$= \sum_{j\in J_\ell} |x_\ell - x_j|\cdot\tilde{u}_{\ell,j}\ .$$

Therefore if $x \in V_0$ and $(x, u_i)_{V_0} = 0$, $i \in I$, then $x=0$. This implies that

(7.1) $\sum_{i \in I} \mathbb{R} u_i = V_0$.

(7.2)**Proposition.** $\sum_{j \in I \setminus \{i\}} {}^tY_iY_i$ *is strictly positive definite for all* $i \in I$.

Proof. Since $\sum_{j \in I} \mathbb{R} u_j = V_0$ and $\dim V_0 \geqq \#(I)-1$, there is an $\ell \in I$ such that $\sum_{j \in I \setminus \{\ell\}} \mathbb{R} u_j = V_0$. Let $V^i = \{x \in V_0;\ Y_j x = 0,\ j \in I \setminus \{i\}\}$, $i \in I$. Then it is easy to see that $\sum_{j \in I \setminus \{i\}} {}^tY_jY_j$ is strictly positive definite if and only if $V^i = \{0\}$.

Note that if $x \in V^\ell$, then $(u_j, x)_{V_0} = s_j^{-1}({}^tY_j u_j, x)_{V_0} = 0$, $j \in I \setminus \{\ell\}$. So we see that $V^\ell = \{0\}$. For each $i \in I$, there is a symmetry U such that $U(\pi(\langle i \rangle)) = \pi(\langle \ell \rangle)$.

Since $U(\{\pi(\langle j \rangle);\ j \in I\}) = \{\pi(\langle j \rangle);\ j \in I\}$, U induces a permutation σ on I. Note that each permutation τ on I induces an isometry T_τ on $\mathbb{R}^I$ by $(T_\tau x)_j = x_{\tau(j)}$, $j \in I$. Noting that F_j, $j \in I$, is a unique 1-cell containing $\pi(\langle j \rangle)$, we see that $U(F_j) = F_{\sigma(j)}$, and so there is a permutation σ_j on I such that $U(\pi(\langle j\ell \rangle)) = \pi(\langle \sigma(j)\sigma_j(\ell) \rangle)$, $\ell \in I$, for each $j \in I$. Therefore we have

$$a_{\ell k}^{(j)} = u_k(\pi(\langle j\ell \rangle)) = u_{\sigma(k)}(U(\pi(\langle j\ell \rangle)))$$
$$= u_{\sigma(k)}(\psi_{\sigma(j)}(\pi(\langle \sigma_j(\ell) \rangle))),\ j, \ell, k \in I.$$

This implies that

$$Y_j = T_{\sigma_j} Y_{\sigma(j)} T_\sigma^{-1},\ j \in I.$$

So we see that if $x \in V^i$, $0 = Y_j x = T_{\sigma_j} Y_{\sigma(j)} T_\sigma^{-1} x$, $j \in I \setminus \{i\}$. This implies that $T_\sigma^{-1} x \in V^\ell$, and so we see that $V^i = \{0\}$.

This completes the proof.

From now on, we assume the following.

Assumption 1. *For each symmetry, there is a permutation* σ_U *on*

$\{0,\dots,N\}$ *such that* $U\psi_k = \psi_{\sigma_U(k)}U$, $k=0,\dots,N$.

Assumption 2. $\sum_{i\in I} \mathbb{R}(Pv_i) = V_0$.

Let $R_{\sigma_U}:\Omega\to\Omega$ be given by $R_{\sigma_U}(\omega)(n) = \sigma_U(\omega(n))$, $n\in\mathbb{N}$, for each symmetry U. Then we see that

(7.3) $U(\pi(\omega)) = \pi(R_{\sigma_U}(\omega))$

for any symmetry U and $\omega\in\Omega$. Let $\mathcal{G}_0 = \{\sigma_U;$ U is a symmetry$\}$. Then we see that $\sigma(I) = I$ for all $\sigma\in\mathcal{G}_0$. So each $\sigma\in\mathcal{G}_0$ induces an isometry T_σ in $\mathbb{R}^I$ given by $(T_\sigma x)_i = x_{\sigma(i)}$, $i\in I$. Observe that for any $k\in\{0,\dots,N\}$,

$$\begin{aligned} a_{ij}^{(k)} &= u_j(\pi(\langle ki\rangle)) = u_{\sigma(j)}(U(\pi(\langle ki\rangle))) \\ &= a_{\sigma(i)\sigma(j)}^{(\sigma(k))}, \quad i,j\in I. \end{aligned}$$

This implies that

(7.4) $Y_k = T_\sigma^{-1}Y_{\sigma(k)}T_\sigma$

for any $\sigma \in \mathcal{G}_0$ and $k\in\{0,1,\dots,N\}$.

Remind that s_i is the second maximum eigen value of Y_i, $i\in I$. Therefore we see that s_i, $i\in I$, are the same. Let us denote it by s_0. Also, $\mathbb{R}\cdot u_i$ and $\mathbb{R}\cdot(Pv_i)$ are the eigen space corresponding to the eigenvalue s_0 of ${}^tY_i$ and Y_i respectively. Therefore we have

$$T_\sigma u_i = u_{\sigma(i)},$$

and

$$T_\sigma v_i = v_{\sigma(i)}$$

for any $i\in I$ and $\sigma\in\mathcal{G}_0$.

(7.5)**Proposition.** (1) *If* V *is a vector space such that* $Y_i(V)\subset V$, $i\in I$, *and* $T_\sigma(V) = V$, $\sigma\in\mathcal{G}_0$, *then* $V = \{0\}$ or V_0.

(2) *If* V *is a vector space such that* ${}^tY_i(V)\subset V$, $i\in I$, *and* $T_\sigma(V) = V$, $\sigma\in\mathcal{G}_0$, *then* $V = \{0\}$ or V_0.

Proof. Since the proofs of the assertions (1) and (2) are similar, we prove only the assertion (1). Suppose that $V \neq \{0\}$. Then since

$\sum_{i\in I} \mathbb{R}u_i = V_0$, we see that there are $i\in I$ and $x\in V$ such that $(u_i,x)_{V_0}\neq 0$. Then we have $(u_i,x)_{V_0}\cdot Pv_i = \lim_{n\to\infty} s_0^{-n}\cdot Y_i^n x \in V$, and so $Pv_i\in V$. Since $T_\sigma Pv_i = Pv_{\sigma(i)}$, $\sigma\in\mathcal{G}_0$, $\{\sigma(i);\ \sigma\in\mathcal{G}_0\} = I$ and $\sum_{j\in I}\mathbb{R}(Pv_j) = V_0$, we see that $V = V_0$.

This completes the proof.

(7.6)**Proposition.** *There is a unique strictly positive symmetric operator* Q_1 *in* V_0 *such that* $\sum_{k=0}^{N} Y_k Q_1 {}^tY_k = (1-c)Q_1$, $T_\sigma^{-1}Q_1T_\sigma = Q_1$, $\sigma\in\mathcal{G}_0$, *and* $\mathrm{trace}(Q_0Q_1) = 1$.

Proof. Let S be a set of non-negative definite symmetric operators in V_0 such that $\mathrm{trace}(QQ_0) = 1$ and $T_\sigma^{-1}QT_\sigma = Q$, $\sigma\in\mathcal{G}_0$. For any symmetric operator Q in V_0, let $F(Q) = (1-c)^{-1}\sum_{k=0}^{N} Y_k Q {}^tY_k$. Then since $\sum_{k=0}^{N} {}^tY_k Q_0 Y_k = (1-c)Q_0$, and $T_\sigma^{-1}Y_kT_\sigma = Y_{\sigma(k)}$, $k=0,\ldots,N$, and ${}^tT_\sigma = T_\sigma^{-1}$, $\sigma\in\mathcal{G}_0$, we see that $F|_S$ is a continuous map from S to S. Since S is a compact convex set, there is a $Q_1 \in S$ such that $F(Q_1) = Q_1$.

Suppose that Q is a non-negative definite symmetric operator in V_0 satisfying $F(Q) = Q$ and $T_\sigma^{-1}QT_\sigma = Q$, $\sigma\in\mathcal{G}_0$. Let $V = \{v\in V_0;\ (Qv,v)_{V_0} = 0\}$. Then it is easy to see that ${}^tY_k(V) \subset V$, $k=0,\ldots,N$ and $T_\sigma(V) = V$, $\sigma\in\mathcal{G}_0$. So by Proposition(7.3) we have $V = \{0\}$ or V_0. This shows that any non-negative definite symmetric operator Q in V_0 satisfying $F(Q) = Q$ is strictly positive definite or zero. So Q_1 is strictly positive definite. Moreovoer, suppose that there are two elements Q_1 and Q_2 in S such that $F(Q_i) = Q_i$, $i=1,2$. Let $a_0 = \max\{a>0;\ Q_1-aQ_2$ is non-negative definite$\}$. Then $F(Q_1-a_0Q_2) = Q_1-a_0Q_2$, and $Q_1-a_0Q_2$ is non-negative definite but not strictly positive definite. So $Q_1=a_0Q_2$. Since $\mathrm{trace}(Q_0Q_1) =$

trace(Q_0Q_2), we have $Q_1=Q_2$. This proves the uniqueness.

This completes the proof.

Let Q_1 be the symmetric operator in V_0 as in Proposition(7.6). Let $\mu = \mu^{(Q_1)}$ be a probability measure in Ω as given in Definition(5.2). Then μ is a stationary measure in Ω. Let S denote the set of nonnegative definite symmetric operators A in V_0 such that trace$(Q_1A) = 1$. Then we see that trace$((\sum_{k=0}^{N} {}^tY_kAY_k)Q_1) = 1-c > 0$. Let $\tilde{S} = \{0,\ldots,N\}\times S$. We define a function $a: S\times\{0,\ldots,N\}\to[0,1-c]$ by

$$a(A,k) = \mathrm{trace}(Q_1\,{}^tY_kAY_k)\ ,\ A\in S,\ k=0,\ldots,N.$$

Now let us define a probability measure $P((k,A),\cdot)$ in $\tilde{S}$ for each $(k,A)\in\tilde{S}$ by

$$(7.7)\quad P((k,A),E) = \sum_{\substack{0\le j\le N\\ a(A,j)\ne 0}} (1-c)^{-1}a(A,j)\cdot\chi_E((j,a(A,j)^{-1}\cdot{}^tY_jAY_j))$$

for any $E\subset\tilde{S}$. Let $\{P_z;\ z\in\tilde{S}\}$ be a family of probability measure in $\Theta = \tilde{S}^{\{0\}\cup\mathbb{N}}$ which defines the Markov chain associated $P(z,\cdot)$, i.e.,

$$P_z[\{\ \theta\in\Theta\ ;\ \theta_0\in E_0,\ \theta_1\in E_1,\ldots,\ \theta_n\in E_n\ \}]$$
$$= \chi_{E_0}(z)\cdot\int_{E_1\times\ldots\times E_n} P(z,dz_1)P(z_1,dz_2)\ldots P(z_{n-1},dz_n)$$

for any $z\in\tilde{S}$, $E_0,E_1,\ldots,E_n\in\mathcal{B}(\tilde{S})$.

For each $\sigma\in\mathcal{G}_0$, let $\tilde{T}_\sigma:\tilde{S}\to\tilde{S}$ be a map given by $\tilde{T}_\sigma(k,A) = (\sigma(k),T_\sigma^{-1}AT_\sigma)$. Then we see that $P(\tilde{T}_\sigma z,\tilde{T}_\sigma^{-1}(E)) = P(z,E)$ for any $z\in\tilde{S}$ and $E\subset\mathcal{B}(\tilde{S})$. Therefore we see that the probability laws of $\tilde{T}_\sigma w(\cdot)$ under $P_z(dw)$ and $w(\cdot)$ under $P_{\sigma(z)}(dw)$ are the same.

Let m_0 be the probability distribution in $\tilde{S}$ induced by $(\omega(1),Z(\omega))$ under μ.

(7.8)**Proposition.** m_0 *is P-invariant*, i.e.,

$$m_0P(E) \underset{\mathrm{def}}{=} \int_{\tilde{S}} m_0(dz)P(z,E) = m_0(E)\ .$$

Proof. By Proposition(5.6)(3), we have
$a(Z(T\omega),\omega(1))^{-1}\cdot {}^tX_1(\omega)Z(T\omega)X_1(\omega) = Z(\omega)$, μ-a.s.ω. Therefore by Proposition(5.6)(6), we have for any $f \in C(\tilde{S};R)$,

$$\begin{aligned}
&\int_{\tilde{S}} f(z)\, m_0(dz) \\
&= E^{\mu}[\, E^{\mu}[f(\omega(1),Z(\omega)))|\mathcal{F}_2^{\infty}]] \\
&= E^{\mu}[\, \int_{\tilde{S}} f(z)P((k,Z(T\omega)),dz)\,]\ ,\ k=0,\dots,N, \\
&= \int_{\tilde{S}} m_0(dz_1)\int_{\tilde{S}} f(z_2)P(z_1,dz_2)\ .
\end{aligned}$$

This proves our assertion.

Let pr_S denotes the natural projection map from $\tilde{S}$ into S. Let $S_p = \{ A \in S;$ the rank of A is p$\}$, p = 1,..., dim V_0 and let $\tilde{S}_p = pr_S^{-1}(S_p)$. Let M denote the set of all P-invariant probability measure m in $\tilde{S}$ such that P_m is ergodic. Since $P(z,\tilde{S}_p) = 1$ if $z \in \tilde{S}_p$, we see that for each m∈M there is a $p \in \{1,\dots,\dim V_0\}$ such that $m(\tilde{S}_p) = 1$.

(7.9)**Proposition.** *For any* $m \in M$, $\{ v \in V_0;\ \sum_{\sigma\in\mathcal{G}_0} (T_\sigma v, AT_\sigma v)_{V_0} = 0$ for all $A \in pr_S(supp(m))\ \} = \{0\}$.

Proof. Let $V = \{ v \in V_0;\ \sum_{\sigma\in\mathcal{G}_0} (T_\sigma v, AT_\sigma v)_{V_0} = 0$ for all $A \in pr_S(supp(m))\ \}$. For $(k,A) \in supp(m)$, $(j, a(A,j)^{-1}\cdot {}^tY_jAY_j) \in supp(m)$, if $a(A,j)\neq 0$, j = 0,...,N. Therefore we see that

$$\sum_{j=0}^{N}\sum_{\sigma\in\mathcal{G}_0} (T_\sigma v, {}^tY_jAY_jT_\sigma v) = \sum_{j=0}^{N}\sum_{\sigma\in\mathcal{G}_0} (T_\sigma Y_j v, AT_\sigma Y_j v) = 0 \text{ for any } v \in V$$

and $A \in pr_S(supp(m))$. So we have $Y_jV \subset V$, j = 1,...,N. Also, we see that $T_\sigma V = V$, $\sigma\in\mathcal{G}_0$. Since $V \neq V_0$, $V = \{0\}$.

This completes the proof.

(7.10)**Proposition.** *Let* $Z_n^{(A)}(\omega) = (\text{trace } Q_1{}^tW_n(\omega)AW_n(\omega))^{-1}\cdot {}^tW_n(\omega)AW_n(\omega)$ *for any* $A \in S$ *and* $\omega \in \Omega$. *Then* *we have*

$$P_{(k,A)}[\ pr_S(\theta_0) \in C_0, \ldots, pr_S(\theta_n) \in C_n\ ]$$
$$= E^{\mu}[\ (\operatorname{trace} Q_1\, {}^tW_nAW_n)(\operatorname{trace} Q_1 Z_n(\omega))^{-1},\ Z_0^{(A)} \in C_0, \ldots, Z_n^{(A)} \in C_n\ ]$$

for any $n \geqq 1$ *and* $C_0, \ldots, C_n \in \mathcal{B}(S)$.

Proof. This comes from the following.

$$P_{(k,A)}[\ pr_S(\theta_0) \in C_0, \ldots, pr_S(\theta_n) \in C_n\ ]$$
$$= \sum_{k_1,\ldots,k_n=0}^{N} (1-c)^{-n}\cdot \operatorname{trace} Q_1\, {}^tY_{k_1} \cdots {}^tY_{k_n} AY_{k_n} \cdots Y_{k_1}$$
$$\cdot \chi_{C_0}(A)\chi_{C_1}(B(A,k_1)) \cdots \chi_{C_n}(B(\ldots B(B(A,k_1),k_2),\ldots,k_n)$$
$$= E^{\nu}[\ \lambda^{-n}\cdot \operatorname{trace} Q_1\, {}^tW_n(\omega)AW_n(\omega),\ Z_\ell^{(A)}(\omega) \in C_\ell \text{ for } \ell = 0,\ldots,n\ ]$$
$$= E^{\mu}[\ (\operatorname{trace} Q_1\, {}^tW_nAW_n)(\operatorname{trace} Q_1 Z_n(\omega))^{-1},\ Z_0^{(A)} \in C_0, \ldots, Z_n^{(A)} \in C_n\ ].$$

Here $B(A,k) = a(A,k)^{-1}\cdot {}^tY_kAY_k$, for $A \in S$ and $k = 0,1,\ldots,N$ with $a(A,k) \neq 0$.

This completes the proof.

(7.11)**Theorem.** *Let* p_0 *is the index of* $((Y_0,\ldots,Y_N),((N+1)^{-1},\ldots,(N+1)^{-1}))$. *Then*

(7.12) $\mu[\ \operatorname{rank} Z(\omega) = p_0\ ] = 1$ and

(7.13) $p_0 = \max\{\ p \in \{1,\ldots,\dim V_0\};\ m(S_p) = 1 \text{ for some } m \in M\ \}$.

Proof. Since $m_0(\tilde{S}_{p_0}) > 0$ by Corollary(5.9), we see that there is an $m \in M$ such that $m(\tilde{S}_{p_0}) = 1$. Let $q = \max\{\ p \in \{1,\ldots,\dim V_0\};\ m(S_p) = 1 \text{ for some } m \in M\ \}$ and take an $m \in M$ such that $m(\tilde{S}_q) = 1$. Then $q \geqq p_0$.

Since P_m is ergodic, we see that

$$P_z[\ \frac{1}{n}\sum_{k=1}^{n} f(\theta_k) \to \int_{\tilde{S}} f\, dm \text{ for all } f \in C(\tilde{S};R)\ ] = 1,\ m\text{-a.e.}z.$$

Therefore, by Proposition(7.9), we see that there are $(k_j,A_j) \in \tilde{S}$, $j = 1,\ldots,d$, such that $\sum_{\sigma\in\mathcal{G}_0}\sum_{j=1}^{d} T_\sigma^{-1}A_jT_\sigma$ is strictly positive definite

and

$$P_{(k_j,A_j)}\left[\ \frac{1}{n}\sum_{k=1}^{n} f(\theta_k) \to \int_{\tilde S} f\,dm \quad \text{for all } f \in C(\tilde S;\mathbf{R})\ \right] = 1$$

for $j = 1,\ldots,d$. Therefore

$$\lim_{n\to\infty} \frac{1}{n}\sum_{k=1}^{n} E^{P_{\tilde T_\sigma(i_j,A_j)}}[f(\theta_k)] = \int_{\tilde S} f\,dm\circ\tilde T_\sigma^{-1}$$

for any $f \in C(\tilde S;\mathbf{R})$, $j = 1,\ldots,d$ and $\sigma\in\mathcal{G}_0$.

Take a $\delta > 0$ such that $\delta\, Q_0 \leqq \sum_{\sigma\in\mathcal{G}_0}\sum_{j=1}^{d} T_\sigma^{-1}A_jT_\sigma \leqq \delta^{-1}Q_0$.

Let $\rho_{j,\sigma,n}(\omega) =$

$(\mathrm{trace}\ Q_1\,{}^tW_n(\omega)T_\sigma^{-1}A_jT_\sigma W_n(\omega))\cdot(\mathrm{trace}\ Q_1\,{}^tW_n(\omega)Q_0W_n(\omega))^{-1}$, $n \geqq 1$, $\sigma\in\mathcal{G}_0$

and $j = 1,\ldots,d$. Then we have

$$Z_n^{(T_\sigma^{-1}A_jT_\sigma)}(\omega) \leqq (\delta\cdot\rho_{j,\sigma,n}(\omega))^{-1}\cdot Z_n(\omega) \quad\text{and}\quad \sum_{\sigma\in\mathcal{G}_0}\sum_{j=1}^{d}\rho_{j,\sigma,n}(\omega) \geqq \delta.$$

Therefore, letting $\tilde d = d\cdot\#(\mathcal{G}_0)$, we have for any $\varepsilon > 0$

$$\mu[\ \|\Lambda^q Z_n\|^q_{\Lambda^q V_0\to\Lambda^q V_0} < \varepsilon\ ]$$

$$\leqq \sum_{\sigma\in\mathcal{G}_0}\sum_{j=1}^{d}\mu[\ \|\Lambda^q Z_n^{(T_\sigma^{-1}A_jT_\sigma)}\|^q_{\Lambda^q V_0\to\Lambda^q V_0} \leqq \tilde d^q\delta^{-2q}\varepsilon,\ \rho_{j,\sigma,n} \geqq \tilde d^{-1}\delta\ ]$$

$$\leqq \tilde d\cdot\delta^{-1}\cdot\sum_{\sigma\in\mathcal{G}_0}\sum_{j=1}^{d} E^{\mu}[\ \rho_{j,\sigma,n},\ \|\Lambda^q Z_n^{(T_\sigma^{-1}A_jT_\sigma)}\|^q_{\Lambda^q V_0\to\Lambda^q V_0} \leqq \tilde d^q\delta^{-2q}\varepsilon\ ]$$

$$\leqq \tilde d\cdot\delta^{-1}\cdot\sum_{\sigma\in\mathcal{G}_0}\sum_{j=1}^{d} P_{\tilde T_\sigma(i_j,A_j)}[\ \|\Lambda^q(pr_S(\theta_n))\|^q_{\Lambda^q V_0\to\Lambda^q V_0} \leqq d^q\delta^{-2q}\varepsilon\ ].$$

Therefore we have

$$\mu[\ \|\Lambda^q Z\|^q_{\Lambda^q V_0\to\Lambda^q V_0} < \varepsilon\ ]$$

$$\leqq \overline{\lim_{n\to\infty}}\frac{1}{n}\sum_{k=1}^{n}\mu[\ \|\Lambda^q Z_k\|^q_{\Lambda^q V_0\to\Lambda^q V_0} < \varepsilon\ ]$$

$$\leqq \tilde d\cdot\delta^{-1}\cdot\sum_{\sigma\in\mathcal{G}_0}\sum_{j=1}^{d}\overline{\lim_{n\to\infty}}\frac{1}{n}\sum_{k=1}^{n} P_{\tilde T_\sigma(i_j,A_j)}[\ \|\Lambda^q(pr_S(\theta_k))\|^q_{\Lambda^q V_0\to\Lambda^q V_0} \leqq d^q\delta^{-2q}\varepsilon\ ]$$

$$\leqq \tilde d^2\delta^{-1}\cdot\sum_{\sigma\in\mathcal{G}_0}(m\circ\tilde T_\sigma^{-1})((\ z \in \tilde S;\ \|\Lambda^q(pr_S(z))\|^q_{\Lambda^q V_0\to\Lambda^q V_0} \leqq d^q\delta^{-2q}\varepsilon\ ))$$

$\to 0$ as $\varepsilon \downarrow 0$.

This proves that $\mu[\ \text{rank } Z(\omega) \geqq q\ ] = 1$. This implies our assertion.

Let $G_p(V_0)$, $p = 1,\ldots,\dim V_0$, be the set of all p-dimensional vector subspace of V_0. Then G_p is a compact manifold. Let φ_p be a map from S_p into G_p defined by $\varphi_p(A) = \{\ Av \in V_0;\ v \in V_0\}$, $A \in S_p$. Then φ_p is a continuous map. Let $\tilde{\varphi}_p = \varphi_p \circ pr_S : \tilde{S}_p \to G_p$.

(7.14)**Theorem.** *Let* q *be the index. Then* $m_0 \circ \tilde{\varphi}_q^{-1}$ *is absolutely continuous relative to* $\sum_{\sigma \in \mathcal{G}_0} m \circ \tilde{T}_\sigma \circ \tilde{\varphi}_q^{-1}$ *for any* $m \in M$ *with* $m(\tilde{S}_q) = 1$. *Moreover, its Radon-Nykodim density is bounded.*

Proof. For any compact set K in G_q and $\varepsilon, \gamma > 0$, let $\bar{K}_{\varepsilon,\gamma}$ be a set given by

$$\bar{K}_{\varepsilon,\gamma} = \{\ A \in S_q;\ \text{there are } A' \in S_q \text{ and } B \in S \text{ such that } \varphi_q(\tilde{A}) \in K,$$
$$\|A'\|_{\overset{q}{\wedge} V_0 \to \overset{q}{\wedge} V_0} \geqq \gamma,\ \|A'-B\|_{V_0 \to V_0} \leqq \varepsilon \text{ and } B \geqq \gamma A\ \}\ .$$

Then $\bar{K}_{\varepsilon,\gamma}$ is a compact set in S and $\bigcap_{\varepsilon > 0} \varphi_q(\bar{K}_{\varepsilon,\gamma}) \subset K$ for any $\gamma > 0$.

Let us use the notion in the proof of Theorem(7.11). Take a $\gamma >$ with $\gamma < \tilde{d}^{-1}\delta^2$. Then we have

$$\sum_{\sigma \in \mathcal{G}_0} (m \cdot T_\sigma^{-1})(pr_S^{-1}(\bar{K}_{\varepsilon,\gamma}))$$
$$\geqq \tilde{d}^{-1} \cdot \sum_{\sigma \in \mathcal{G}_0} \sum_{j=1}^{d} \overline{\lim_{n\to\infty}} \frac{1}{n} \sum_{k=1}^{n} P_{\tilde{T}_\sigma^{-1}(i_j, A_j)}[\ \theta_k \in \bar{K}_{\varepsilon,\gamma}\ ]$$
$$\geqq \tilde{d}^{-1} \cdot \overline{\lim_{n\to\infty}} \frac{1}{n} \sum_{\sigma \in \mathcal{G}_0} \sum_{j=1}^{d} \sum_{k=1}^{n} E^{\mu}[\ \rho_{j,\sigma,k} \cdot Z_k^{(T_\sigma^{-1} A_j T_\sigma)} \in \bar{K}_{\varepsilon,\gamma}\ ]$$
$$\geqq \tilde{d}^{-1} \cdot \overline{\lim_{n\to\infty}} \frac{1}{n} \sum_{\sigma \in \mathcal{G}_0} \sum_{j=1}^{d} \sum_{k=1}^{n} \tilde{d}^{-1}\delta \cdot \mu[\ \rho_{j,\sigma,k} \geqq \tilde{d}^{-1}\delta,\ \|Z-Z_k\|_{V_0 \to V_0} \leqq \varepsilon,$$
$$\|\overset{q}{\wedge} Z\|_{\overset{q}{\wedge} V_0 \to \overset{q}{\wedge} V_0} \geqq \gamma\ ,\ \varphi_q(Z) \in K\ ]$$

$$\geqq \tilde{d}^{-2}\delta\cdot \overline{\lim_{n\to\infty}} \frac{1}{n}\sum_{k=1}^{n} \mu[\ \|Z-Z_k\|_{V_0\to V_0} \leqq \varepsilon,\ \|\overset{q}{\Lambda} Z\|^q_{\overset{q}{\Lambda}V_0\to\overset{q}{\Lambda}V_0} \geqq \gamma,\ \varphi_q(Z)\in K\ ]$$

$$= \tilde{d}^{-2}\delta\cdot \mu[\ \|\overset{q}{\Lambda} Z\|^q_{\overset{q}{\Lambda}V_0\to\overset{q}{\Lambda}V_0} \geqq \gamma,\ \varphi_q(Z)\in K\ ]\ .$$

Letting $\varepsilon \downarrow 0$ first and letting $\gamma \downarrow 0$, we have

$$m_0\circ\tilde{\varphi}_q^{-1}(K) \leqq \tilde{d}^2\delta^{-1} \sum_{\sigma\in\mathcal{G}_0} m\circ\tilde{T}_\sigma^{-1}\circ\tilde{\varphi}_q^{-1}(K)$$

for any compact set K in G_q.

This implies our assertion.

(7.15)**Corollary.** *Suppose that the index equals one. Let* $m\in M$. *Then* $M = \{m\circ T_\sigma^{-1};\ \sigma\in\mathcal{G}_0\}$. *Moreover, there is an ergodic probability measure* μ_0 *on* Ω, *and* $\mu = (\#(\mathcal{G}_0))^{-1}\cdot \sum_{\sigma\in\mathcal{G}_0} \mu_0\circ R_\sigma^{-1}$. Here $R_\sigma:\Omega\to\Omega$ is a map given by $R_\sigma\omega(n) = \sigma(\omega(n))$, $n\geqq 1$, for each $\sigma\in\mathcal{G}_0$.

Proof. Note that $\varphi_1:S_1\to G_1(V_0)$ is one-to-one. Therefore we see that $m_0\circ pr_S^{-1}$ is absolutely comtinuous relative to $\sum_{\sigma\in\mathcal{G}_0} (m\circ\tilde{T}_\sigma^{-1})\circ pr_S^{-1}$ for any $m\in M$. Since $P((A,k),E)$, $E\in\mathcal{B}(\tilde{S})$ is independent of k, we see that $m_0 = m_0P$ is absolutely continuous relative to $\sum_{\sigma\in\mathcal{G}_0} m\circ\tilde{T}_\sigma^{-1} = \sum_{\sigma\in\mathcal{G}_0} (m\circ\tilde{T}_\sigma^{-1})P$. Therefore P_{m_0} is absolutely continuous relative to $\sum_{\sigma\in\mathcal{G}_0} P_{m\circ\tilde{T}_\sigma^{-1}}$. But this implies that $m_0 = (\#(\mathcal{G}_0))^{-1}\cdot \sum_{\sigma\in\mathcal{G}_0} m\circ\tilde{T}_\sigma^{-1}$.

This completes the proof.

(7.16)**Corollary.** *If the index is one, then* $\mu=\nu$, *or* μ *and* ν *are mutually singular.*

Proof. Let μ_0 be the ergodic measure on Ω as in Corollary(7.15). Then $\nu = \mu_0$, or ν and μ_0 are mutually singular. Since $\nu\circ R_\sigma = \nu$ for all $\sigma\in\mathcal{G}_0$, we have our assertion.

References.

[1] Barlow, M.T., and R.F. Bass, The construction of Brownian motion on the Sierpinski carpet, Ann. Inst. H. Poincaré, Prob. et Stat. 25(1989), 225-258.

[2] Barlow, M.T., and R.F. Bass, Local times for Brownian motion on the Sierpinski carpet, to appear in Prob. Theo. Rel. Fields.

[3] Barlow, M.T., R.F. Bass, and J. D. Sherwood, Resistence and spectral dimension of Sierpinski carpets, Preprint.

[4] Barlow, M.T., and E.A. Perkins, Brownian motion on the Sierpinski gasket, Prob. Theo. Rel. Fields 79(1988), 543-624.

[5] Falconer, K.J., The geometry of fractal sets, Cambridge University Press, Cambridge, 1985.

[6] Feder, J., Fractals, Plenum Publishing Company, New York-London, 1988.

[7] Fukushima, M., Dirichlet forms and Markov Processes, North-Holland/Kodansha, Amsterdam/Tokyo, 1980.

[8] Fukushima, M., and Shima T., On a spectral analysis for the Sierpinski gasket, Preprint.

[9] Goldstein, S., Random walks and diffusions on fractals, Percolation theory and ergodic theory of infinite particle systems (Minneapolis, Minn., 1984-85), pp. 121-129, IMA Vol. Math. Appl. 8, Springer, New York-Berlin, 1987.

[10] Hattori, K., T. Hattori, and S. Kusuoka, Self-avoiding random walk on the Sierpinski gasket, Prob. Theo. Rel. Fields 84(1990), 1-26.

[11] Hattori, K., T. Hattori, and H. Watanabe, Gaussian fields theories on general networks and the spectral dimensions, Progress Theo. Physics Suppl. 92(1987), 108-143.

[12] Hutchinson: Fractals and self-similarity, Indiana Univ. Math. J. 30(1981), 713-747.

[13] Kigami, J., A harmonic calculus on the Sirpinski spaces,

Japan J. Appl. Math. 6(1989), 259-290.

[14] Kusuoka, S., A diffusion process on a fractal, Probabilistic methods in Mathematical Physics, Proc. of Taniguchi International Symp. (Katata and Kyoto, 1985) ed. K.Ito and N.Ikeda, pp. 251-274, Kinokuniya, Tokyo, 1987.

[15] Kusuoka, S., Dirichlet forms on fractals and product of random matrices, Publ. RIMS Kyoto Univ. 25(1989), 659-680.

[16] Linstrom, T., Brownian motion on nested fractals, Preprint.

[17] Mandelbrot, B.B., The fractal geometry of nature, W.H. Freeman and Co., San Fransisco, 1983.

[18] Osada, H., Isoperimetric dimension and estimates of heat kernels of pre-Sierpinski carpet, to appear in Prob. Theo. Rel. Fields.

[19] Rammal, R., and G. Toulouse, Random walks on fractal structures and percolation clusters, J. Physique Lettres 44(1983), L13-L22.

[20] Shima T., On eigenvalue problems for the Sierpinski pre-gaskets, to appear in Japan J. Appl. Math.

R.I.M.S.
Kyoto University